The People's Army

The People's Army

THE HOME GUARD IN SCOTLAND 1940–44

Brian D. Osborne

BIRLINN

First published in 2009 by
Birlinn Limited
West Newington House
10 Newington Road
Edinburgh
EH9 1QS

www.birlinn.co.uk

ISBN: 978 1 84341 043 0

British Library Cataloguing-in-Publication Data
A catalogue record for this book is available from the British Library

Designed and typeset by SJC
Printed and bound by CPI Cox & Wyman, Reading

CONTENTS

ILLUSTRATIONS

ACKNOWLEDGEMENTS

I would like to express my thanks to those people who have made photographs and other illustrations available for this book; they are individually acknowledged in the credits to the illustrations. I am also particularly grateful to Bill McChlery and James Walker of Kirkintilloch for helpful and interesting conversations about their experiences in the Home Guard.

I am of course indebted to librarians and archivists for access to books and manuscripts dealing with the Home Guard. In particular it is a pleasure to acknowledge the contribution of the National Library of Scotland which holds an invaluable set of unit histories assembled by a far-seeing army officer at the end of the Second World War. Quotations from these manuscripts are made with the permission of the Trustees of the National Library of Scotland. The National Archives of the United Kingdom at Kew have rich resources dealing with the Home Guard in Scotland and quotations from these documents are made with the permission of the National Archives. The Home Guard was the subject of much debate in both houses of parliament and parliamentary material is reproduced with the permission of the Controller of HMSO on behalf of parliament.

Chapter 1

IMAGES OF THE HOME GUARD

More than sixty years after the Home Guard stood down it is perhaps not surprising that the image of the organisation in most British minds is formed as much by the popular BBC television series *Dad's Army*, first broadcast in 1968–77, as by first- or second-hand memories of the organisation or by any published accounts of its role and history. The Home Guard has of course been the subject of a number of books, indeed a semi-official account by Charles Graves, *The Home Guard of Britain*, was published in 1943 while the war was still raging and John Brophy's paean to the movement, *Britain's Home Guard: A Character Study*, illustrated by Eric Kennington's heroic and idealised images of Home Guardsmen, appeared in 1945. More recently S.P. MacKenzie's scholarly and invaluable *The Home Guard: A Military and Political History* was published in 1995 and local and unit histories are also to be found. However there has not previously been an attempt to look at the Home Guard in Scotland as a whole, despite its significance and impact: around 155,000 Scotsmen (and some women) were enrolled in the organisation when it stood down in 1944 and about 250,000 in all passed through its ranks.

The Home Guard is a curiously difficult organisation to pin down; its function changed several times in the four and a half years of its life and some of the inherent contradictions within it were never entirely resolved. What is certainly true is that it was a much more complicated organisation and one that was militarily, socially and politically far more significant and interesting than might at first be thought if the *Dad's Army* image was to be taken as the full story.

Studying contemporary documents and press reports about the Home Guard (or, as it was called for the first few months of its life, the Local Defence Volunteers) one is frequently reminded of the Walmington-on-Sea platoon depicted in the television series. The authors of the series, David Croft and Jimmy Perry, were often extremely accurate in the historical context for their comedy. Yes, there were any number of blimpish Captain Mainwarings who rose to command; yes, there were many veterans of Queen Victoria's wars who, like Lance-Corporal Jones, lied about their age to enlist and to stay enrolled; yes, there were ineffectual youths like Private Pike (named after the Home Guard's least favourite weapon, the bayonet standard or pike issued in 1942); and yes, the initial formation of the Local Defence Volunteers was just as much a scene of chaos and improvisation as the series portrayed it to be.

What is perhaps less clearly appreciated, if one takes *Dad's Army* as normative, are the very considerable tensions within the organisation. Tensions were perhaps predictable between the establishment figures who emerged to occupy the senior ranks and those who saw the LDV or the Home Guard as a "People's Army", interpreted as a left-wing anti-fascist militia with echoes of the International Brigade in the Spanish Civil War. There were also tensions between quite senior Home Guard officers and regular army officers; the Home Guard often felt that some regular officers issuing orders from a remote headquarters did not understand the dynamics of leading and controlling an unpaid, part-time, largely voluntary movement, a perception which was probably often accurate. It is also fair to note that at least the higher command of the army went to considerable lengths to keep the Home Guard happy and committed. There were also tensions between the original volunteers and the men who, after 1941, were drafted into the Home Guard – drafted to serve without pay, of course. There were tensions between the War Office and the awkward child it had fathered, tensions which were largely caused by the independent and voluntary spirit of the movement – a classic example being the formation of an unauthorised but extremely popular training school for the LDV in July 1940 at Osterley Park near London. This school, the brainchild of Tom Wintringham, the former commander of the British battalion of the International Brigade, was not only looked upon with suspicion because of the left-wing nature of many of the staff, but because it was energetically promoted in the popular magazine *Picture Post*. Wintringham proved to be unacceptable as a Home Guard officer because of his Communist Party background and several others of the Osterley Park instructors were prevented from joining the Home Guard because of their left-wing leanings. There was in place a general ban on persons with either fascist or communist connections joining the LDV or the Home Guard. Wintringham wrote in *Picture Post*: "My own view, the only explanation that I can yet see for these facts, is that there is a reactionary clique with some power in the War Office, at least one of whom actively assisted German and Italian tanks, planes and troops to conquer Spain. They prefer that Home Guard training should be delayed, rather than that soldiers who fought in Spain against the German tanks and planes, and against Italian troops, should be recognised by the Home Guard as useful men to follow."[1]

Paradoxically, Wintringham, who had been expelled from the Communist Party because of his personal life, was criticised from the left for his enthusiastic support for what Communist Party orthodoxy, up to the German invasion of the Soviet Union, saw as an unjust capitalist war. The Communist Party of Great Britain considered that the German-Soviet Non-Aggression Treaty, or Molotov-Ribbentrop Pact, signed on 23 August 1939 superseded their innate anti-fascist principles. This was a view which they deftly managed to reverse when Hitler launched Operation Barbarossa – the invasion of the Soviet Union – in June 1941 and the Soviet Union became embroiled in what it then described as the Great Patriotic War.

As Wintringham's articles in *Picture Post* suggest, the press swiftly took up the whole LDV/Home Guard movement and ran regular Home Guard columns and features, which were often in their tone critical of bureaucracy and the "establishment". This represented a form of publicity that must have been alien to a War Office used to the innate discretion and orderly and disciplined subordination of a professional army. In addition to newspapers and magazines there was soon a substantial literature of commercially published instructional books covering everything the Home Guardsman might need to know, from camouflage to weapons training and from fieldcraft to medical care, together with more political or theoretical works such as Wintringham's *Armies of Freemen* and *New Ways of War*. All these supplemented a plethora of War Office instructional booklets and leaflets and they presumably sold in significant numbers to the enthusiastic members of the force. A number of authors who were serving Home Guardsmen, such as John Langdon-Davis, an anti-fascist journalist who had covered the Spanish Civil War, contributed to this outpouring of books, as well as writing for newspapers and magazines. The result was that Home Guard policies and practice were freely and publicly debated in a way that must have seemed very strange, if not positively insubordinate, to the traditional military mind.

The LDV and the Home Guard was a mass voluntary movement set up on the basis of equality of service; originally there were no ranks, simply appointments as section or platoon leaders, and so on. Even after it became "militarised" after November 1940 by the introduction of commissions, ranks and saluting it retained strong elements of this ethos of equality; death or disablement pensions, for example, were the same for all ranks. The Home Guard also retained a quite special sense of its own uniqueness and a considerable political power and influence. In a way that was certainly not shared by the regular armed forces of the Crown, it was able and was quite prepared to use whatever channels of influence in parliament and the press that were available to it to fight its case and argue its cause. Its political influence was assisted by the involvement of many politicians in the Home Guard – for example the Palace of Westminster unit counted nine members of the House of Lords and sixty-four members of parliament in its ranks – a vocal and well-placed pressure group in itself. Many other parliamentarians served in other London units or in units in their home area.

It must be remembered that the parliament elected at the general election of 1936 was only 18 years removed from the end of the First World War. The majority of MPs and peers would have had personal military experience in the First World War and would, in consequence, be interested in and feel qualified to discuss military matters. Among the Scottish parliamentarians who regularly asked questions or took part in debates on the LDV or Home Guard were the Earl of Breadalbane, who served in the Westminster LDV and initiated a debate in the House of Lords on 4 June 1940, the Earl of Mansfield, a company commander in the 2nd Perthshire Battalion, and Lieutenant-Colonel Sir Thomas Moore, the Unionist MP for Ayr Burghs, who served in a London battalion of the Home Guard and was a regular speaker on Home Guard matters. Many Scottish MPs were to be active in the force, such as Lieutenant-Colonel Sir John Colville, Unionist MP for Midlothian and Peebles North and Secretary of State for Scotland in the Chamberlain government until April 1940. On going out of office Colville became an LDV area commander and would, in August 1942, be appointed staff officer in charge of Home Guard Liaison at Scottish Command. Nor were those interested in the Home Guard confined to the Conservative & Unionist benches: Robert Gibson, the advocate who represented Greenock in the Labour interest, was keenly interested in the problems of the Home Guard and James Henderson Stewart, Liberal MP for East Fife, served in his local unit in Surrey. Of these parliamentarians Breadalbane had been a gunnery officer on the Western Front in the Great War and had later commanded a Territorial Army battalion of the Argylls, Mansfield had been a lieutenant in a Territorial Army battalion of the Black Watch, Colville had served with the Cameronians (Scottish Rifles), Gibson had been a captain in the Royal Garrison Artillery in the First War, Stewart had been a captain in the Royal Artillery in the First War while Moore had been a regular officer from 1908 to 1925.

The birth of the movement undoubtedly owed much to the age-old political desire to be seen to be doing something rather than nothing. In this case there was a need to address the twin problems of preventing a widespread outbreak of freelance local amateur defence groups and assuaging public fears about the dangers of paratroop landings and fifth-columnists, a point colourfully made by the Independent Labour MP Colonel Josiah Wedgwood in the first major House of Commons Debate on the LDV. He said, "I do not believe that the War Office originally formed this force in order to resist invasion. I believe it was formed because the demand of the people in the country was overwhelming

– the demand to be allowed to have some part, some weapons and some chance to stand up to these devils when they come."[2]

If this was true in May 1940 then the Home Guard was to move from being a sort of semi-armed special constabulary – arguably little more than a public relations gesture or a morale-boosting enterprise – to a serious second-line military force. Such a development could not happen overnight, as Tom Wintringham, writing in the summer of 1940 about his vision of an army to fight a "People's War" observed: "The Local Defence Volunteers are the beginnings of such a force. They have been given too little to do and often the wrong things to do; their organisation and leadership is not yet that of a People's Army. But the force is growing and developing; it can grow until the real eagerness of our people to defend their homes finds full expression within it."[3]

As the threat of invasion receded, with German attention turning to the Eastern Front, so the Home Guard's role expanded from the *Dad's Army* Walmington-on-Sea platoon model, guarding the invasion coast from a strongpoint at the Novelty Rock Emporium, into a variety of other forms of service, most notably in anti-aircraft defence. By so doing it released many tens of thousands of regular troops for service in the field armies.

An organisation which its founders, in their most optimistic speculations, had thought might eventually attract between 100,000 and 200,000 men ended up with a peak strength of just under 2,000,000, surely a "People's Army" in strength and in its widespread influence, if not in the Marxist revolutionary sense.

Chapter 2

THE CRISIS OF 1940

Prime Minister Neville Chamberlain broadcast to the nation at 11.15 a.m. on Sunday 3 September 1939. He told the British people that assurances had been sought from the German government that the German invasion of Poland, which had been launched on 1 September, would be halted. Sir Neville Henderson, the British Ambassador in Berlin, had delivered an ultimatum to the German Foreign Ministry at 9 a.m. that morning. Chamberlain told his listeners that no response had been received by the deadline of 11.00 a.m. and that, "consequently this country is at war with Germany."

Britain was committed, in terms of her pre-war agreements with France, to send an expeditionary force to the Continent. Chamberlain indeed announced in his speech that: "We and France are today, in fulfilment of our obligations, going to the aid of Poland. . ." How exactly this aid was to be manifested in practical terms was not made clear; indeed when Chamberlain was asked on 7 September in the House of Commons whether it would be possible for the Allies to undertake some military operations that might relieve pressure on Poland he declined to answer.

The British army in 1939 consisted of two elements, a professional regular force and the Territorial Army, a part-time volunteer force that would be mobilised for full-time duty in an emergency. The Territorial Army, or TA, was made up of weekend soldiers based in local drill halls up and down the country. It would form an essential component of the British Expeditionary Force (BEF) sent to the Continent and was organised, armed and equipped on the same lines as the regular army. Many TA units had indeed been converted from their traditional cavalry or infantry roles to meet the changed demands of modern war. Thus the 9th (Dunbartonshire) Battalion of the Argyll & Sutherland Highlanders lost its infantry role in 1938 and three of its four companies became the 54th (Argyll & Sutherland Highlanders) Light Anti-Aircraft Regiment of the Royal Artillery. The balance of the battalion went into the 54th Anti-Tank Regiment which had been formed from the Queen's Own Royal Glasgow Yeomanry, a TA cavalry regiment which had been converted to field artillery back in 1922.

Territorial Army units were intended to stand alongside regular army units in the line of battle and were integrated with such units; indeed on mobilisation the Territorial Army as such ceased to exist and became an integral part of the army. Of the infantry battalions making up the BEF on 10 May 1940, the day of the German attack in the west, 49 were regular units and 70 came from what had been the Territorial Army. The distinction between the two is still worth noting at this stage of the war, not least for the local identity of many of the units. For example, the 152nd Infantry Brigade of the 51st (Highland) Division, a Territorial Army formation, which was positioned in May 1940 in front of the French fortified position known as the Maginot Line, comprised one regular battalion, the 2nd Seaforth Highlanders, and two Territorial Battalions, the 4th Seaforth Highlanders, recruited from Easter Ross and the 4th Queens Own Cameron Highlanders, recruited from Inverness-shire and Nairn. The 54th

Light Anti-Aircraft Regiment found itself in May 1940 attached to III Corps of the BEF.

The Territorial Army had been doubled in size in March 1939, and each battalion had been ordered to split itself in half and create a duplicate unit. This was to some extent a paper increase as the availability of men and materials to sustain these additional units was not always apparent. In many cases one of the new units was seen as the first-line force and the other – with older, lower medical category men – was seen as a reserve force for home defence.

The Military Training Act of June 1939 registered all men in the 20 to 21 age group for six months' military training – the first time compulsory military service had been introduced in Britain in peacetime. On the outbreak of war the National Service (Armed Forces) Act introduced conscription for men aged between 18 and 41. Of course neither of these measures meant that every man in those age groups was immediately called up; some had their service deferred or set aside because they were employed in essential industries, and there was in any event a limit to the training personnel and facilities available to the armed forces. Men were registered in age groups, the 40-year-old group only registering in June 1941, a fact of considerable significance for the number of younger men who were thus free to enrol, at least temporarily, in the LDV or the Home Guard.

The BEF took the cream of the British army, regular and TA, to France and Belgium, although there remained substantial forces located in overseas bases in India, the Middle East and in various colonial territories. Although initially four divisions were sent across in September 1939, by the beginning of May 1940 this had risen to ten full-strength infantry divisions and a tank brigade, plus three under-strength, untrained and under-equipped divisions in training or support roles which did not form part of the front line of the BEF. Other units were forming in Britain and were sent to France after the German attack.

The forces remaining in the United Kingdom were stripped of equipment to ensure that the BEF was armed as adequately as possible. Had the BEF been destroyed or captured in France and Belgium there would have been only a very small defensive force remaining in Britain to deal with an invasion, even though Home Forces were reinforced by the arrival of Canadian troops, the first echelon of which landed at Greenock on 25 December 1939. The fact that the bulk of the personnel of the BEF were successfully evacuated back to Britain from Dunkirk and elsewhere strengthened the numbers of troops available for the defence of the United Kingdom, but as most of their heavy equipment had been abandoned on the Continent there was a critical shortage of arms and equipment, a shortage which could not be quickly rectified – a fact that would have repercussions on the establishment, arming and training of the Local Defence Volunteers and the Home Guard.

Once the Territorial Army had been mobilised on 1 September 1939, and thus integrated into the regular army structure, there was no effective reserve of part-time volunteers to call on. A very half-hearted attempt had been made in 1936 to create National Defence Companies; these were to be recruited from among ex-servicemen who might be able to defend vulnerable points in wartime, and replaced the moribund Royal Defence Corps which had been formed in 1917 from garrison battalions of eighteen infantry regiments. However, the Companies were never properly funded, hardly existed as more than paper units, and never reached their extremely modest target strength of 8,450 officers and men. Even the outbreak of war failed to breathe much life into this organisation, and recruitment, other than for officers, was officially halted in October 1939. The formation of replacement units, the Home

Defence Battalions, was announced, although even these were not on a very ambitious scale with a supplementary recruitment target of just 20,000 ex-servicemen aged between 35 and 50, hardly a large number when dispersed over the length and breadth of Britain.

The period between September 1939 and the German attacks on Denmark and Norway in April 1940 became known as the Phoney War; a phrase which was never remotely applicable to the war at sea but did sum up the situation on land and initially in the air – which was perhaps the only effective area where the Allies could have given direct aid to Poland. At the outbreak of war Britain and France had resolved not to bomb anything but the most narrowly defined military targets. Quite how narrowly defined these were soon emerged.

The Conservative MP Leo Amery told how he learned that the RAF was restricted to dropping propaganda leaflets on Germany. He took this matter up with the Secretary of State for Air, Sir Kingsley Wood, on 5 September and suggested that, as Germany was reported to be short of timber, the RAF should attempt to set fire to the Black Forest. This proposal was rejected out of hand: "To my consternation he told me that there was no question of our bombing even the munitions works at Essen which were private property, or lines of communication, and that doing so would alienate American opinion. To my question whether we were not going to lift a finger to help the Poles he had no answer."[1] The fact that the suggested targets were private property rather than the possessions of the German state put them off-limits in Wood's view; the concept of total war was evidently taking a little time to be appreciated in some quarters.

The BEF was not to be tested until 10 May when the German army bypassed the much-vaunted French defensive position of the Maginot Line and, in a plan devised by Erich von Manstein, swept through Belgium and Holland. The invasion of the Netherlands and Belgium was marked by the effective use of airborne troops to seize key points and cause dislocation to the defences.

German success in Norway had been aided by the existence of a local fascist movement, the Nasjonal Samling Party under Vidkun Quisling. Quisling, an ex-officer of the Norwegian army, was in Berlin in April 1940 and helped plan the invasion which was intended to set him up as the head of a puppet government.

These two novel elements; the threat of landings of airborne troops by parachute, glider or conventional powered aircraft, and fear of an indigenous "fifth column" of fascist sympathisers and potential collaborators, or, as they were swiftly named, "Quislings," were to be powerfully influential forces behind the creation of what became Britain's citizen army.

Only seventeen days elapsed between the German *blitzkrieg* in the west and the start of the evacuation from Dunkirk. A French and British intervention in Norway had proved to be ineffective and the attempt by Neville Chamberlain to reconstruct his government on a national coalition basis failed because of a lack of confidence in the Liberal and Labour ranks, and also among many Conservatives, in Chamberlain's leadership. In a stormy session of the House of Commons on 7 May that had debated the Norwegian campaign one of these dissident Conservatives, Leo Amery, called for: ". . .a real National Government. Somehow or other we must get in the Government men who could match our enemies in fighting spirit, daring, resolution and thirst for victory. . . We could not go on being led as we are."[2] Chamberlain resigned on 10 May and was replaced by Winston Churchill, whom he had recalled from years on the backbenches on 3 September 1939 to be First Lord of the Admiralty.

The government may not at this stage have taken the need for local home defence entirely seriously but up and down the country local groups had recognised the need and had started to organise their own system of local defence. In Port William in Wigtownshire, Dr Gavin Brown, a local general practitioner, organised 120 volunteers who took turns, from dusk to dawn, to man four coast-watching points and a telephone reporting base at the local police station. Such passive defence might have been acceptable to the authorities but in other areas armed groups were patrolling in search of German parachutists.

This type of freelance activity was looked upon with a considerable degree of alarm by the civil and military powers; the prospect of bands of armed civilians roaming the countryside was seen as a highly undesirable development. To be fair to the authorities such concerns were well founded. There was, to start with, a problem in international law. Irregular fighters who did not wear uniform would be considered as *franc-tireurs* (literally free-shooters) and as such fell outside the protection of the Geneva Convention and were liable, if captured, to be executed rather than treated as prisoners of war. It was, of course, a legalistic rather than a moral distinction, and the novelist G.K. Chesterton had summed it up pithily in 1919: "In other words, a *franc-tireur* is you or I or any other healthy man who found himself, when attacked, in accidental possession of a gun or pistol, and not in accidental possession of a particular cap or a particular pair of trousers."[3]

In July 1940, after the establishment of the Local Defence Volunteers, the German media issued warnings that what it described as "armed gangs" of civilians would be treated as murderers if they took up arms against German soldiers. Bremen Radio, broadcasting in English, was reported by *The Times* to have reminded its listeners of the execution of Polish *franc-tireurs* and to have said that the "preparations which are being made all over Britain to arm the civilian population for guerrilla warfare are contrary to the rules of international law"[4] The British government's view was expressed by Sir Edward Grigg, MP, the Under Secretary of State for War, who had day-to-day responsibility for the LDV/Home Guard. In a broadcast on 15 June he stated: "...under the Regulation by which you were established your Local Defence Volunteers are part of the Armed Forces of the Crown. You, therefore, rank as soldiers, with a soldier's rights, and a soldier's obligations. The most important of your rights is to use armed force against the enemies of your country, the most vital of your obligations is unquestioning obedience to your leaders."[5]

Much of the *franc-tireurs* argument centred on uniform and whether, as the British government claimed, a brassard or armband with the legend LDV qualified as a uniform. A Labour MP, E.N. Bennett, writing to *The Times*[6] expressed doubts and quoted the relevant passage of the Hague Convention which called for "*un signe distinctif fixé et reconnaisable à distance*" and recalled that in the First World War Germany had refused to accept a brassard, by itself, as conferring belligerent rights as it could easily be slipped on and off and was thus not "*fixé*." He also questioned whether an armband was in fact recognisable at a distance. In fact the War Office instructions on the armband, which was in any event intended as a temporary measure until uniforms of one sort or another were made available, had been that it should be sewn on the outer garment. The Germans returned to the topic at the end of July and a Berlin evening paper, reported in *The Times*, claimed that: "The British Government is evolving a type of warfare in which weapons will be carried openly in the civilian's hands and in which individual men and women will threaten the lives of German soldiers from ambush with every conceivable kind of treachery."[7] This concern from the German state-controlled press is indicative both of a degree of concern about the threat posed by the Home Guard and of the official German view that an invasion of Britain was a foregone conclusion.

However, quite apart from such legal concerns about the niceties of the Hague Convention and international law regarding *franc-tireurs* there were of course perfectly sound reasons why military commanders would not want a third force with no command, control and communication links to their headquarters intervening in a possible combat situation. Steps were taken by the War Office to consider how to manage this development and to find out exactly what was happening, and urgent messages went out along the lines of the following one sent to the Air Raid Precautions Controller in Fife at 01:00 hours on 13 May by the District Commissioner for Civil Defence in Dundee: "Unconfirmed report received in London says that bands of civilians are forming all over country and arming themselves with shot-guns etc for the purpose of detecting and dealing with German parachutists. Will you please report the position in your District (Fife) to District Commissioner's Office by 0800 hours on 13th May 1940."[8] The Fife ARP Controller was able to reassure the authorities that the Kingdom of Fife, at least, was free from roaming bands of shotgun toting civilians.

The anti-parachutist role adopted by these bands of volunteers, and which also formed part of the initial *raison d'être* of the LDV, was in fact predicated on a false assumption. It was popularly imagined that German airborne troops, or air-dropped saboteurs, would drift down slowly and allow what the press delighted to call "parashots" or "parashooters" ample time to pick them off in mid-air with a well-aimed blast from a 12-bore shotgun. Indeed the Air Ministry helpfully pointed out on 10 May that, as no British aircraft had a crew of more than six, any group of more than six parachutists that might be seen were not bailing out of a stricken aircraft but could be considered as enemy combatants and, as such, fair targets. However, as Hugh Slater wrote in 1941: "After about a fortnight it was realised that the Eschner parachute, used by the Germans, is designed so that it may take no longer than five seconds for the parachutists to be landed from the plane, and therefore the concept of potting at them as they floated through the air had to be abandoned as obsolete."[9]

By 14 May the military situation on the Continent was grave. Norway and the Netherlands had set up governments in exile in the United Kingdom. The press reported that British and French forces were rushing to the support of the retreating Belgian forces and that the German army had broken through the first Dutch defence line on the River Ijsell.

At home Anthony Eden, who had been Secretary of State for Dominion Affairs in Chamberlain's government, had been appointed Secretary of State for War in Churchill's coalition government as a replacement for Oliver Stanley. While these political changes were taking place a series of meetings were being held in the War Office, initially attended by Oliver Stanley, but driven forward by the desire of the Commander in Chief Home Forces, General Sir Walter Kirke, to channel the outburst of civilian enthusiasm for military affairs into more organised, acceptable and controllable lines. One of the rejected ideas was for the attachment of volunteers to anti-aircraft searchlight batteries, which were the most widespread regular military units in the country, and largely static formations.

However within two days the outline of the final successful scheme had been drawn up. Drawn up, it must be confessed, in a remarkably loose and casual way. The officers responsible had, for lack of time, bypassed the normal Civil Service channels and failed to take the proper legal, financial and administrative advice, a failure which later had serious consequences for the administration of the new force. A few weeks after the formation of the Local Defence Volunteers Major H. Broun Lindsay was moved to write to the *Scotsman* newspaper: "Having been entrusted with the task of organising the county of East Lothian against the landing of parachutists, I am at once faced with the need for money for the provision of armbands and ammunition other than .303. I have already incurred an expenditure of roughly £50 on these necessaries. As time is a vital factor I felt justified in getting these materials on my own responsibility."[10] He concluded by appealing for donations to defray his costs and ended, "The Government provides nothing except a number of rifles and .303 ammunition." The unfortunate Major Broun Lindsay's financial problem, and as £50 in 1940 was equivalent to around £1,800 in purchasing power today it was a fairly substantial one, was clearly a consequence of the haste and improvisation which had marked the establishment of the new force.

A senior civil servant pointed out with some acerbity to the military officers who were producing the plan that the creation of the Territorial Army had taken years of planning and an act of parliament. They were told that they were setting up a major military organisation on the basis of a couple of meetings and two days' work and that financial and legal difficulties were to be expected as a result.

Eventually the outline scheme for the creation of a force, which was to be called the Local Defence Volunteers, had been finalised and the new secretary of state advised of the arrangements. Through him cabinet approval was obtained for the plan. The army's intention had been for General Kirke to broadcast a radio appeal for volunteers but Eden felt that, as Secretary of State, it was more appropriate that he, and not the Commander in Chief Home Forces, should speak to the nation; doubtless as the newly appointed war minister he saw this as an opportunity to establish himself with the public. A broadcast on the BBC Home Service was arranged for the evening of 14 May 1940. As Eden prepared to speak the situation on the Continent was deteriorating by the hour: a major tank battle had been fought with French losses the previous day; the Belgian city of Liège had fallen; heavy parachute landings were reported from the Netherlands; the Germans were approaching the river Meuse on their advance through the Ardennes; Queen Wilhelmina of the Netherlands arrived in Britain as a refugee; Rotterdam was bombed, and the Netherlands Commander in Chief would be obliged to sign instruments of surrender on 15 May.

Chapter 3

THE LOCAL DEFENCE VOLUNTEERS

At 18:20 hours on 14 May, just three hours before Eden's speech, district commissioners, central government's local representatives dealing with civil defence and related matters, were advised of the impending broadcast and of the formation of a new defence force to deal with parachutists. Although the secretary of state's broadcast would instruct volunteers to report to their local police station to register for the new force, there was not time for the information to filter down to every police station or indeed for any forms to be distributed. The hope was that by the 15th the police would have become aware of the plans to establish the new force and be ready to undertake their limited, but essential role, as the registration agency. However, as Anthony Eden records in his memoirs, the first recruit reported to a doubtless greatly surprised police officer within four minutes of the broadcast ending. A lawyer was the first Edinburgh volunteer, turning up at a city police station shortly after the broadcast.

In his speech Eden outlined the use the Germans had made of parachute troops in Holland and Belgium: "The purpose of the parachute attack is to disorganise and confuse, as a preparation for the landing of troops by aircraft." His listeners, who would have represented a very large percentage of the UK population who would be tuned into the 9 o'clock Home Service news bulletin at this critical stage in the war, were told that the success of any such airborne attack depended on speed, and that although enemy aircraft would have to penetrate the anti-aircraft defences of the country it was important to supplement existing defences with a new and locally based one. Eden told how the government since 1939 had been inundated with requests from men who were, for one reason or another, not engaged in military service and who wished to contribute in some way to the national defence. He felt sure the new body he was announcing would be welcomed by those previously frustrated volunteers. He said: "We want large numbers of such men in Great Britain, who are British subjects, between the ages of 17 and 65 to come forward now and offer their service in order to make assurance doubly sure. The name of the new force which is now to be raised will be the Local Defence Volunteers. This name, Local Defence Volunteers, describes its duties in three words. It must be understood that this is, so to speak, a spare-time job, so there will be no need for any volunteer to abandon his present occupation."[1]

Eden's reference to Great Britain was deliberate and specific, as there was no plan to extend the Local Defence Volunteers to Northern Ireland. In view of the strength of the religious and political divisions in the island of Ireland it was considered unwise to establish an armed civilian force there. Ulster had a devolved parliament and it later decided to raise an analogous body, the Ulster Defence Volunteers, but placed it under the control of the overwhelmingly Protestant and Unionist Royal Ulster Constabulary. By so doing it made the volunteer force less attractive to the Catholic and Nationalist part of the population.

Eden went on to say that part-time members of existing civil defence organisations should seek advice from their officers before registering and that younger men who would eventually be called up under the National Service (Armed Forces) Act could join the Volunteers temporarily and be released when their call-up papers arrived. He continued:

> Now a word to those who propose to volunteer. When on duty you will form part of the armed forces, and your period of service will be for the duration of the war. You will not be paid, but you will receive uniform and you will be armed. You will be entrusted with certain vital duties, for which reasonable fitness and a knowledge of firearms are necessary. These duties will not require you to live away from your homes.
>
> In order to volunteer what you have to do is to give your name at your local police station; and then, as and when we want you, we will let you know.

The secretary of state emphasised that his appeal was mainly directed to those in small towns, villages, and less densely populated urban areas – the belief apparently being that parachutists were not likely to land in the major centres of population. This initial unwillingness to recruit in large towns and cities was quickly and quietly dropped in the face of overwhelming public pressure to volunteer in these areas; the first, though by no means the last, occasion on which popular sentiment would affect the policy of the force. He concluded: "Here, then, is the opportunity for which so many of you have been waiting. Your loyal help, added to the arrangements which already exist, will make and keep our country safe."

On 21 May, Eden, answering questions in the House of Commons on the Local Defence Volunteers, gave a little more detail and observed that although he did not have authoritative figures for recruiting, the response to his appeal had been most satisfactory. He went on:

No establishment has been fixed, and the numbers accepted will depend on the circumstances in each area. . .
All volunteers will be enrolled as soldiers and there will be
no officers or non-commissioned officers in the ordinary
army sense of those terms, nor will there be any pay or
other emoluments. Compensation will, however, be given
for injuries attributable to service. Services with the Force
will be for the duration of the emergency, unless a man
is in the meantime called up under the National Service
(Armed Forces) Act, but may be terminated earlier either
by the competent authority at any time or by the Volunteer himself on giving a fortnight's notice. The Force will be
supplied with arms, ammunition and uniform.[2]

Whatever Eden's expectation for the new force, and he later
said that his own hope had been that it might attract between
100,000 and 200,000 men across Britain, he and the War Office can hardly have believed the scale of response. The *Glasgow
Herald* of 16 May reported that around 1,200 men had called at
city police stations on the 15th to register, and in Edinburgh a
squad of women auxiliary police had to be drafted in to deal with
an almost overwhelming rush of applications for the new force. A
similar story was reported from around Scotland. Chief Constable Smith of Hamilton Burgh Constabulary reported an enthusiastic response: "Everybody around here who is capable of holding
a gun wants to have a pop at the parachutists." In Stonehaven 83
volunteers enrolled and over 100 gave in their names in Hawick.
The *Scotsman* of 16 May reported that in Edinburgh volunteers
probably ran into the thousands and were drawn from all ranks
of society: "Civil Servants, University lecturers, students, joiners,
plumbers, scavengers and labourers." Despite the advertised age
limit of 65 there were many men of 70 and 75 registering and

the *Glasgow Herald* on 17 May noted that many of the older men
registering in Edinburgh "were careful to state their qualifications
as 'crack shot'", doubtless hoping this would improve their chances
of being enrolled.

By 17 May the War Office was reporting over 250,000 applicants and every part of the country was sharing in the enthusiasm.
In Aberdeen city over 600 men had enrolled by the afternoon of
the 16th and across Scotland ex-servicemen and senior school-
boys rubbed shoulders with rifle club members and gamekeepers
in the queues to enrol.

The Earl of Elgin was later to write in an account of the Home
Guard in Fife and Kinross that: "The women were most bitterly
disappointed that they too were not allowed to shoulder arms if
need be."[3] The position of women in the Local Defence Volunteers was to be a contentious issue and their enrolment was officially barred, although in April 1943 women were at last allowed
to become Home Guard auxiliaries and permitted to undertake
clerical, catering, communications and orderly room duties. In
fact many women had been undertaking these duties from the
early days of the force. In June 1940 Miss Mary Forsyth, clerkess
at the Macallan whisky distillery, volunteered to do the typing and
correspondence for the headquarters of the 1st Banffshire Battalion, a portable typewriter having been loaned for this purpose
by Miss Marshall of the Highland Hotel, Craigellachie. The 1st
Banffshire could also draw on the services of one male and three
female teachers at Buckie High School as German interpreters
– a useful resource if German prisoners had to be questioned.

However these support roles did not go far enough to satisfy
many women, most notably Dr Edith Summerskill, Labour MP
for Fulham West, who campaigned vigorously for the enrolment
of women as full members of the Home Guard. Speaking in an adjournment debate on the Home Guard in the House of Commons

on 19 November 1940, after pointing to the heavy farm work being done by members of the Women's Land Army "...big, hefty, healthy girls with strong nerves...", she said: "One must get rid of this idea that women are still weak, gentle creatures who must be protected. Hon. Members may think that, but the modern enemy does not. Why, therefore, if women are treated this way, should not they also be allowed to defend themselves? Let us cast aside all prejudices and dismiss the 19th century conception of womanhood. Let us recognise the sterling work of our women in this battle, and give them a chance to join this Home Guard so that they may defend their own country."[4]

Dr Summerskill's campaign failed to win women the right to carry a rifle in the LDV or the Home Guard but her sentiments were echoed by many women. Dorothy Reid, writing from Busby, Renfrewshire, to the *Glasgow Herald* argued that women could usefully assist the LDV by providing day-time spotters – a role which, she pointed out, women had carried out in Finland in the Finno-Russian war. A female letter-writer from Argyllshire asked: "Why should not we women living in the wilder and lonelier parts of the country – such as the Western Highlands of Scotland – be also armed and organised for the defence of our homes and farms? Perhaps the Government might see its way to include women in the new Local Defence Volunteer Force."[5]

The terms of the secretary of state's broadcast could not have been clearer; men between the ages of 17 and 65 were what were wanted, but in the outpouring of patriotic enthusiasm these limits were widely ignored. The *Glasgow Herald* of 23 May reported that ex-Provost Harvey of Greenock had asked to be enrolled in the Volunteers, despite being 80 years of age. Doubtless many other men over 65 conveniently misplaced their birth certificates and assured the desk officer at their police station that they were mere youths of 64. The restriction on over 65-year-olds serving was one that attracted much discussion. Although later it was more rigorously enforced, it was never quite as hard and fast a rule as officialdom might have wished. In a parliamentary question on 17 September 1940 Anthony Eden was asked by the MP for Greenock, Robert Gibson, how many applicants under 65 had failed on physical grounds to satisfy the medical examiners and had not been enrolled and how many men over 65 had passed the physical examination but had been refused on grounds of age. He replied:

> Mr Eden: There is no medical examination of applicants for enrolment in the Home Guard. As regards the age limit, I would refer my Hon. and learned Friend to the answer which I gave him on 13th August.
> Mr Gibson: Will not my right Hon Friend reconsider this matter? Am I to gather from the first part of his answer that it is now possible for men over 65 to enrol in the Home Guard?
> Mr Eden: I think the Hon. and learned Gentleman is assuming much too much.[6]

Mr Gibson may have been assuming too much but in fact the presence in the Home Guard of quite elderly men was winked at. The 3rd Glasgow Battalion counted among the ranks of its machine-gun company Sergeant J. Stewart, who died in service in 1944 at the age of 76. Sergeant Stewart had enlisted in the army as a drummer boy in 1882 and had served in Egypt, South Africa and during the First World War. He had volunteered for the Home Guard in 1940 at the age of 72 – or seven years over the official age limit – and had obviously continued to serve for more than four years.[7]

Colonel Josiah Wedgwood, Independent Labour MP for Newcastle under Lyme, an enthusiast for the LDV and the Home

Guard who became a regular contributor to debates on the movement, speaking on an Adjournment Debate in the House on 22 May 1940, made the case for the value of older volunteers: "Every retired military man up to the age of 80, and every officer, has joined this Force, and put his age down as something else. These people are well qualified to do this work without going dithering about like sheep, as will happen to inexperienced troops, asking somebody else what they ought to do."[8] Colonel Wedgwood had a personal interest; aged 68, he had registered for the LDV but had at the time of the debate heard nothing more about his enlistment.

On 15 May, the day following Eden's broadcast, the War Office started to lay down the rules for the organisation of the new force. A message was sent to all home commands setting out the organisational structure. At each of the home commands, such as Scotland or Southern Command, there would be General Staff Officer Grade II (GSO II) – a major's appointment – to deal with LDV matters and at area HQs, such as Glasgow, Edinburgh and Highland there would be a GSOIII – a post of captain's rank – to handle the LDV. Each area HQ would also have an unpaid area organiser.

Below the area level sat zones, with an unpaid voluntary zone commander. Beneath the zone was the group – again commanded by an unpaid volunteer. The army area commander in consultation with the lords lieutenant of the counties in his area would appoint a suitable person to act as LDV area organiser, zone and group organisers. Very roughly speaking, zones equated to counties. Thus Lanarkshire was originally designated as Zone II of Glasgow area and was divided into two groups, each of which was further subdivided into a number of geographically coherent subgroups such as Hamilton, Blantyre and Bothwell, which formed Sub Group V of Group II. In less populous parts of the country several counties might be linked, thus Wigtownshire, Kirkcudbright and Dumfries were combined into Zone IV of Glasgow area.

The city of Glasgow itself, with the adjacent counties of Renfrew and Dumbarton, formed Zone I. Within the city the decision was taken to recruit and organise in 10 sectors, corresponding to the police divisions of the city.

Many of these arrangements would, in the light of experience, have to be altered, and, particularly at the higher level, there followed an often confusing series of reorganisations. In fairness these initial dispositions were made when the force was expected to number 150,000 to 200,000 across the whole of Britain and had to be modified in light of the vastly greater numbers recruited. For example what was to become the 3rd Dunbartonshire Battalion of the Home Guard started out in life as the Bearsden sub-group of the Dunbartonshire LDV Group, recruiting in the Clydebank, Bearsden and Milngavie areas. Later it was designated the Bearsden Company of the 1st Dunbartonshire Battalion of the Home Guard. By September 1940 this unit was so large that the Clydebank men were detached and formed into the 2nd Dunbartonshire while the Bearsden and Milngavie volunteers, together with those from Kirkintilloch and the Cumbernauld area, were established as the 3rd Dunbartonshire Battalion. As numbers grew the Kirkintilloch and Cumbernauld men were later split off to create the 4th Dunbartonshire Battalion, leaving the Bearsden and Milngavie men constituting the 3rd Battalion.

However, some of the organisational changes were undoubtedly due to an inadequate appreciation of local structures, boundaries and loyalties. The Earl of Breadalbane, speaking in a debate he initiated in the House of Lords on 4 June 1940, complained that Scotland had been divided by an arbitrary line and that "this line only appeared on military maps, and therefore not one member of the civilian force would have a map to show it." He continued, "an amendment was sent out, and another line drawn which cuts across watersheds, lines of communication and other matters of strategic importance."[9]

Area commanders would advise zone and group organisers of the number of volunteers to be raised in each area. This figure would be based on the number of rifles to be allotted to the area, the extent of weapon sharing that would be desirable and the number of other weapons that were available locally. Group organisers would then select men for enlistment from the lists of volunteers who had registered at local police stations. The qualifications were restated:

a) They must be men between the ages of 17 and 65
b) They must be British subjects
c) They must be of reasonable physical fitness[10]

All these three criteria were to cause problems over the life of the force, especially the linked issues of age and physical fitness, although eventually a solution would be found with the Home Guard developing the capacity to find different roles for older and less fit men in static defence posts and for the younger men in attack and battle platoons.

The War Office telegram emphasised that volunteers would be unpaid, would not receive rations and that denim overalls (not normal army battledress) and field service caps would be provided on loan to personnel while on duty.

Across Scotland the new force was taking shape in what was an extraordinarily short period of time. At the centre, the General Officer Commanding Scotland (GOC), General Carrington, recruited Lieutenant-Colonel Wynne, Secretary of the Edinburgh Territorial Army Association, on 16 May to assist in the organisation of the new force and to take up the post of GSO II (LDV)

at Scottish Command. The next day Carrington organised a conference of his area commanders and an LDV area organiser was
appointed for each of his areas, with group organisers at county
level. Nothing more was heard of the idea that the LDV would
not recruit in cities and large towns, the political impossibility of
turning away tens of thousands of eager volunteers having impressed itself on the authorities. Glasgow would have enrolled
14,393 men in nine battalions of the LDV by 7 July. It was infi-
nitely better to have these men inside the organisation than en-
gaging in freelance vigilante activities outside, even if this meant a
swift re-evaluation of the LDV's role.

While these command arrangements were being made, the
force on the ground was taking shape and local initiative was to
the fore. In an account of what was to become the 4th Perthshire
Battalion, Lieutenant-Colonel J.M.D. MacKenzie wrote that he
was asked to raise a section at Balbeggie, near Scone, and 40 to 50
men enrolled and immediately started night patrols: "I went out
myself with Mr Brewster and Mr Douglas. We were armed with
one of my rifles, a sporting .256, a gun and a .22, as well as sticks.
We met a patrol from either Errol or Glencarse, led by Mr Gilroy
of Fingask round about Dalreichmoor, stalked each other hope-
fully in the moonlight, and were rather disappointed that neither
of us was the promised parachutist."[11]

On 18 May the War Office issued a further and more detailed
set of instructions to home commands. This explained that:

2. This Force will form part of the Armed Forces of the
Crown and will be subject to Military Law. But the intention of the Army Council is that for the purposes of administration, the outstanding features should be simplicity, elasticity and decentralised control, coupled with the
minimum of regulations and formalities.

3. There will be no establishment for the Force. There will be no officers or non-commissioned officers in the ordinary Army sense of these terms. There will be no pay or other emoluments. The engagement of the volunteer will be for a period not exceeding the duration of the present emergency, but may be terminated by the competent authority at any time, or at his own request on giving fourteen days notice in writing.[12]

Accompanying this letter was a sample enrolment form with the advice that supplies of the form were being sent to area HQs for onward distribution.

The pious hope for a minimum of regulation and formality was perhaps as much governed by the knowledge that the local units would have little or no administrative and clerical back-up as it was by a sense of the urgency of the national crisis. However, government paperwork has a curious habit of being generated despite the best of intentions and Lieutenant-Colonel Moore, MP for Ayr Burghs, speaking in the Commons on 10 November 1940 about the plight of Home Guard battalion commanders, told the House of Home Guard Circular 25 in which Paragraph 3 "... sets out that there are seven Army forms and 25 Army books, and a number of other odds and ends to be kept by one man, without any paid assistance – until the announcement today – with no machinery, no staff, with a business of his own probably."[13]

The enrolment forms that were sent out were based on the original modest anticipation of numbers. What became the 5th Ayr Battalion was sent 300 forms and when they requested a further 1,500 forms were told this would be impossible to supply – so locally printed forms were produced and by January 1941 the unit had a strength of 2,033 in all ranks.

The War Office letter said: "The numbers to be enrolled will

depend entirely upon the numbers available and the numbers required", but clearly the availability of weapons was a significant
determining factor.

Spare weapons and equipment were in short supply; the ex-
pansion of the regular army had meant that reserves even of basic
things like rifles and steel helmets were dangerously low and there
simply were not a quarter of a million .303 rifles in the stores. An
allocation of weapons was made but if rifles were scarce, ammunition was even scarcer; the Skye company of the 2nd Inverness-
shire Battalion in July 1940 had enlisted 567 men who shared 89
.303 rifles and 5 rounds of ammunition per rifle.

Some enterprising units managed to supplement the government issue from less orthodox sources. The 5th Glasgow for example, "... discovered that the Clyde Trust had 50 ancient French
Alpini rifles in their store. These had been left at Glasgow Docks
by mistake. Application was at once made, and the Clyde Trust
willingly granted their use. The rifles were transported day by day,
by private motorcar from Unit to Unit for purposes of Drilling,
etc."[14]

The lack of weapons was, of course, a serious blow to the
morale of the members of the newly formed force who had been
promised uniforms and weapons. An LDV volunteer in a letter to
the *Scotsman* newspaper, signed himself "Shooter" and concluded:
"A man does not mind taking his chance of being killed, but he
doesn't like being made a fool of. If the LDV are not intended to
fight, then do not arm them at all, but if they are to fight, then
they must have such uniforms as will prevent them being treated
as *franc-tireurs*, and such arms and equipment as will enable them
to meet, as they must do, the flower of the German army."[15]

The letter from "Shooter" had been provoked by an official appeal for shotguns. The War Office had ruled that shotguns and
sporting cartridges were legitimate weapons with which to arm

the LDV and the GOC Scotland had appealed for the loan of shotguns and all available cartridges to meet the need for weapons. Some areas naturally were better supplied with shotguns and sporting rifles than others; Skye managed to supplement its 89 government rifles with 20 to 30 privately owned rifles and it will be recalled that Lieutenant-Colonel MacKenzie's first patrol at Balbeggie was armed with two sporting rifles and a shotgun.

The standard rifle of the British army was the SMLE – Short Magazine Lee Enfield – a .303 calibre weapon. Also available were supplies of First World War vintage Canadian-built Ross Rifles and P.14 rifles, a US-built version of the Enfield, both also .303 calibre weapons. However these weapons were not available in quantities nearly sufficient to meet the needs of the rapidly growing LDV, whose number had reached half a million by mid-June. There was much official backing for the merits of the 12-bore shotgun, especially when combined with a ball cartridge, and many of the books for the new force gave handy hints on how to melt down the lead shot from a standard sporting cartridge to make a single large slug. Tom Wintringham, writing in the *Picture Post*, expressed the views of many: "Your weapon may be a tin can of explosive or a shotgun that will only hit at 50 yards. Treasure it until you have a good chance to kill a German. Even if you only get one you have helped beat Hitler."[16]

Despite such official and unofficial enthusiasm for the merits of the shotgun there is no doubt that most LDV volunteers felt that unless they had a rifle of their own they were not "proper soldiers" and, sensitive to morale considerations, the government arranged to buy 500,000 American Springfield rifles. These weapons had lain in US arsenals since 1918 or earlier and were packed with heavy grease, which needed to be removed before the weapon could be used. The Springfield rifle was a .300 calibre weapon and so ammunition was not interchangeable between them and the

British-pattern rifles. Ammunition supply for the .300 rifles was short, and much irritation was expressed over the limited practice opportunities that existed to use the weapons. As the American rifles became available the SMLE, P.14 and Ross rifles were withdrawn, although some perceptive commanders, like the Earl of Stair, the Wigtownshire Zone Commander, insisted on retaining ten .303 rifles per battalion, for training purposes, because of the better ammunition supply position for that calibre.

Even when rifles were available and materials for degreasing and cleaning them were supplied there was, at times, a slight lack of imagination on the part of the military authorities. In November 1940 Lothian zone was at last provided with oil for rifle cleaning purposes; unfortunately this arrived in the form of four 50-gallon barrels with no taps to draw off the oil, nor receptacles to allow it to be divided up among the many battalions in the zone.

Much was made at the time, and much has since been said, of the "old boys' network" that ran the LDV and the Home Guard and certainly at the group and zone command levels the list of organisers appointed did rather resemble extracts from *Who's Who* or *Burke's Peerage*. This is perhaps not surprising, as the involvement of the lords lieutenant of the counties in the selection process would tend to result in men from a similar social background to the lord lieutenant being selected. Just as with members of parliament, where most had served in the 1914–18 War, most of the local "establishment" figures had either been professional soldiers or had served in the Great War and it was hardly probable that area commanders and lords lieutenants would see past retired generals, brigadiers and colonels when making these initial appointments. And nor did they. The Glasgow area organiser was a retired colonel in the Territorial Army, and his three zone commanders were a lieutenant-colonel and two colonels. If the concern about the risks from Fifth Columnists and Quislings –

the whole "enemy within" fear which had arisen from the German seizure of Norway and Denmark and the attack on the Netherlands and Belgium – is kept in mind, it is perhaps less surprising that these key appointments were made from people whom the appointers knew, understood and trusted, from what were seen as "our sort of people," rather than that they should have cast their net wider, into less familiar waters.

A letter to the *Scotsman* dated 16 May 1940 clearly expressed the concerns of many: "Mr Eden's scheme is most welcome, and I do not wonder at the zeal of your correspondents. It should not be overlooked, however, that without extreme care secret agents of the enemy and Fifth Column men may effect entry into the Local Defence Volunteer Force to its confusion, and that its own uniform and credentials may be forged and used fictitiously by audacious invaders. This has been experienced in other countries. The arming of all and sundry would be a very doubtful and risky procedure."[17] In practical terms too, a retired colonel, a pillar of the Territorial Army Association, perhaps involved in business and social affairs in his area was, arguably, likely to be better able to deal with varied and difficult relationships with local authorities, civil defence organisations and the regular services which were inevitable in setting up and successfully establishing a new military force, than a younger man without military experience or an ex-International Brigader of the Spanish Civil War, however energetic one might be or how committed to the anti-fascist cause the other might be.

Although the senior posts of zone area and group organisers were established and filled from the ranks of the good and the great there was the significant point that they were titled "Organiser" not "Commander". Partly this was to avoid any danger of LDV personnel being thought able or entitled to command regular forces and partly because of the surprisingly egalitarian ethos of the new body and its ruling principle of equality of service. This doctrine was defended by Anthony Eden in parliament in August 1940 when he emphasised that the Home Guard (as it was by then called), "...was organised on the basis of equality of service and status. There was, accordingly, no system of rank, but there are appointments suitably graded to the various formations."[18] This democratic system was widely disliked, not least by many of the old soldiers in its ranks who would have found no problems with saying "Sir" and saluting superiors, and it would soon be substantially modified to resemble a normal military force more closely.

Zones, areas and groups were doubtless needed but the Local Defence Volunteers as a movement was built from the grass-roots up. The basic unit was not a battalion but the section located in a specific locality – a reasonable decision in light of the initial objective of the LDV to be a locally based watch and ward force and the lack of a transport and support infrastructure to permit the deployment of large formations. The War Office letter of 18 May emphasised this structure: "Volunteers, when enrolled, will be formed into sections. The normal approximate size of a section (which will be the basic unit) will be ten men, but rigid adherence to this figure is unnecessary. Sections will be grouped into platoons and platoons into companies, according to local defence requirements."[19]

How this formation of units and appointment of leaders worked out in practice is revealed in a number of the histories of the Home Guard compiled after stand-down in 1944 for Scottish Command and deposited in the National Library of Scotland.[20] For example, the history of the 1st City of Edinburgh Battalion relates how a meeting was called by the chairman of the Edinburgh Territorial Army Association. Normally the lord lieutenant might have convened this type of meeting, but in the four Scottish cities the post of lord lieutenant was held by the lord provost of the day, so in Edinburgh the Territorial Army Association took the initiative. Glasgow's Lord Provost Patrick Dollan was to be more actively involved in the raising of the LDV in his city.

It was agreed that the city should be divided into districts for defence purposes along the lines of police districts and that a company should be raised in each district. Lieutenant-Colonel D.A. Foulis was asked to organise No. 1 Company covering the Cramond to Murrayfield area. Foulis had served during the First War and had risen to command of the 10th Cameronians in 1918.

As a first step advertisements were placed in the local papers inviting people with military experience and able to serve as leaders to call for interview. By 23 May Foulis had a list of potential leaders and the next day he and five potential platoon leaders met in his house to discuss raising the company. It is interesting to note, in the light of criticism of "old boys' networks," that Lieutenant-Colonel Foulis emphasises that, with one exception, the platoon leaders were unknown to him and to each other. The next day Foulis and his platoon leaders toured the area to identify potential accommodation for each platoon.

In addition to recruitment through the police stations public meetings were held at schools and golf courses in the company area and a whole platoon was raised from the members of Murrayfield Golf Club. A meeting was arranged at the Royal Burgess Golf Club of volunteers able to provide motorcars or motorcycles; there were certain advantages in raising a company in the affluent western outskirts of Edinburgh.

Each platoon was paraded on Tuesday 4 June for the purpose of nominating section leaders and dividing the men into sections which were ideally to be made up of friends living in the same locality. "Section Leaders on appointment were warned that they would be on probation for one month and would be replaced if

they had proved incompetent or excessively strict."[21] A section
leader was what the army would have called a sergeant. From 4
June the company manned all its defence posts nightly, although
a guard had been set at a local foundry and on Cramond Bridge
before this date.

A slightly different picture emerges from the account of the
raising of the East Lothian LDV by Major Humphrey Broun
Lindsay, an ex-regular officer and former Unionist MP for Glasgow Partick, whose appeal for funds we encountered in Chapter
1. In his history of the 2nd Battalion East Lothian Home Guard
he comments on the problems arising from the hasty formation
of the LDV, the shortage of accommodation caused by the area
already having a large armed forces presence and the lack of planning for the selection of leaders. He wrote:

> [The] best that could be done was to tour the County and
> select a man in each district and appoint him leader without
> regard to the size of his unit. These were mainly composed
> of school-masters in villages of any size, and farmers in the
> country districts, most of whom had soldiered in the 1914–
> 18 war. Materials were purchased for armbands and enrol-
> ment forms were run off on a Roneo. The armbands were
> cut and machined by the CO's wife and other helpers at the
> rate of 250 a day – blue for Company Commanders, green
> for platoon commanders, white for the rank and file. . . .
>
> Companies were based on the Burghs and large vil-
> lages regardless of numbers. Any large nucleus was called
> a Platoon, and the small knots round farms became sec-
> tions. Organisation was not helped by the direct order that
> there were to be no officers, no discipline, no saluting. In
> fact nothing was to be done that would weld the volunteers
> together in the shortest time available.[22]

The composition of the new LDV units obviously varied greatly across Scotland although in most areas a very large number
of veterans of the First World War were recruited and proved
to be vital in training and developing a military atmosphere in
what was, at first, a very unmilitary organisation whose only outward symbol was a brassard with the letters LDV attached to a
civilian jacket. However not every area had as many First World
War veterans as might be expected; rural areas where most of
the population worked in agriculture and where many man had
been in reserved occupations in 1914–18 proved to be less rich in
this resource than urban areas. The 1st Dumfriesshire Battalion's
history notes that most of their recruits were in this category, so
had little experience of drill, weapons and fighting, although there
were more ex-servicemen in the small towns such as Biggar and
Lockerbie. In compensation the countryman often had some experience of handling a shotgun and was perhaps hardier and fitter
than the average town-dweller.

Units recruiting in mining areas or areas of heavy engineering had a younger age profile than average because such workers
were not generally called up for National Service and commanders found that their recruits had a very high standard of physical
fitness even if this was to be taxed by the double burden of long
working hours and Home Guard duties. The account of the 6th
Lanarkshire Battalion explains:

> . . . almost the whole unit were iron and steel and con-
> struction workers or connected with these trades. Most
> of these men were working at most arduous jobs of the
> utmost importance to the National effort and long hours
> and overtime was a heavy strain on their physique. It is not
> surprising therefore that many of the older men took the
> opportunity of retiring when the force was thrown open to

large numbers of directed men. The greater number, however, stuck to their task and even in the last stages of the
movement one could see bus loads after a day of field firing
lasting all Sunday speeding back to Motherwell to change
and go on to the evening shift.[23]

In the Highlands there were obvious problems of distance
and remoteness with groups (the LDV equivalent of a battalion)
stretching from coast to coast, and beyond. Colonel Sir Donald
Cameron of Lochiel was appointed as Group Organiser for No.
III Group, in No. I Zone of the North Highland Sub-Area of
Highland Area. No. III Group was, in plain language, Inverness-
shire, and Lochiel established his headquarters in a TA drill hall
in Margaret Street, Inverness. Brigadier A.D. MacPherson was
appointed as Assistant Organiser and sub-group organisers were
appointed for the Inverness-shire islands of Harris, North Uist,
South Uist and Barra, the distances involved making more direct
command problematic. Four companies were formed: one based
on Inverness and the adjacent country, one based in Badenoch
and the land to the east of Loch Ness, one in Lochaber including Fort Augustus and Kinlochleven and one in Skye. By 1 July
the strength of the LDV in the county had reached 2,500 and
a month later this number had risen to 3,598 volunteers. There
were, however, a mere 1,269 service rifles to share among them.

The group had to take account of the normal issues of defence
against German invaders and saboteurs and the protection of vital communications assets such as the road and rail bridges over
the Findhorn and the Spey. The group's area also included the
strategically important aluminium factories at Fort William and
Kinlochleven. At an early stage these sites were protected from air
attack by LDV-manned Bren guns and Vickers machine-guns. In
the main, these weapons were provided, not through normal army

channels, but by the Ministry of Aircraft Production (MAP); the output of these factories was of course essential to the building of modern aircraft. Lord Beaverbrook, the energetic Canadian press baron Churchill had appointed to run the ministry, considered the defence of his factories a high priority and used many unorthodox means to obtain scarce weapons and equipment to bolster their defence, often to the irritation of others. Many MAP factory LDV/Home Guard units found themselves equipped with Beaverette armoured cars, which would have been useful tools in many areas but were not always appropriately deployed. For example, the Barr and Stroud optical factory in Anniesland, Glasgow, which as well as rangefinders and other optical equipment made bomb sights and gun sights for aircraft, and thus came under the aegis of the MAP, was allocated a Beaverette which dutifully patrolled the streets of Anniesland.

Within two months of Eden's announcement, recruitment to the Local Defence Volunteers had exceeded one million. On 24 July he announced that 1.3 million had been recruited and that, except in districts where targets had not been met, there would be a temporary suspension of recruitment. Essentially this was to allow time for supplies to catch up with manpower.

Reflecting on these events from more than 60 years on it is perhaps hard to appreciate fully the thoughts and motivations of the volunteers. We now know that the German invasion never came, and that historians have argued, with some force, that a German invasion would have been unlikely to succeed while the Royal Navy remained intact and while the Royal Air Force was able to contest mastery of the skies. By studying the Allied invasions of Europe from Sicily to Normandy we can learn what complex operations amphibious landings were and what elaborate preparations were needed to ensure success. Certainly we know that the Germans in 1940 or 1941 did not have the landing craft, the amphibious tanks, or the range of special equipment that the Allies developed in the course of the war. However this is hindsight.

The reality of 1940 was well expressed by the Under Secretary of State at the War Office, Sir Edward Grigg, speaking in a debate on 22 May 1940: "We are living under conditions when imminent peril may descend on us from the sky at any moment, in the dawn to-morrow or in the dusk tomorrow-night."[24]

In that summer of 1940 men flocked to enrol in the Local Defence Volunteers against a backdrop first of all of the German breakthrough in the west, the withdrawal of the British Expeditionary Force to the Channel, the miraculous rescue of the bulk of the BEF (albeit without most of its heavy armament and equipment), the fall of France, and with Britain the lone European power still resisting Hitler's Germany. All the available evidence suggests that these men had very good reason to expect to be coming to grips with German invaders in the fields and streets of Britain in a matter of days or weeks. All the available evidence suggests that the Germans were, for their part, perfectly serious in planning an invasion, provided that the Royal Air Force could be neutralised.

The most formidable military force in the world – an army and air force which had conquered Poland, Denmark, Norway, Belgium, the Netherlands and France – was parked on the Channel coast. It had demonstrated a new way of making war; the rapid attack of mechanised forces supported by air power known as *blitzkrieg*. The huge French army and the elaborate defences of the Maginot Line had been broken or bypassed.

It was this force that 1.3 million men had enlisted in the LDV to resist. To resist with a woefully inadequate supply of rifles, shotguns and ammunition, with, if nothing else was available, pitchforks and blunderbusses. To resist a well-trained, well-equipped, battle-hardened enemy which had scattered all its previous foes. Yet still they enlisted.

Lieutenant-Colonel A.D. Hunter, writing the history of the 6th Perthshire Battalion of the Home Guard, tells of the first recruits to the LDV in the city of Perth and of the wide range of backgrounds and occupations they came from. However, he went on to say:

> ...the greater part of those who responded to the call were men who had fighting experience in 1914–18. Many of those had said in moments of bitterness and disillusion-ment that never again would they take up arms for their country. Reports had led the country generally to believe that the Germans of 1940 were better trained, better equipped, tougher and even more ruthless than counter-parts of the first Great War. The veterans of 1914–18 did not altogether believe this, but they did believe that they would almost certainly be called upon to meet a very for-midable enemy, who so far in the war had proved invinci-ble, and that they would have to face and kill him under handicap of insufficient arms, very little training and none of the benefits of being a cohesive force. Not least, they had known war at first hand and knew all war's savagery and terror. If reports were to be believed they were volun-teering to go into a veritable shambles where the chance of individual survival in a force such as this looked small indeed. Nevertheless, these ex-soldiers, to their everlasting credit, came forward in some cases eagerly and in all cases determinedly.[25]

Chapter 4

A NEW IDENTITY

The steady stream of recruits to the LDV far outstripped the resources available to arm and equip them. Ten days after Eden's broadcast an initial allocation of P.14 rifles was made in Scottish Command, the bulk of them going to Edinburgh and South Highland areas which were seen as the most vulnerable parts of Scotland. However only 10 rounds per rifle were allocated to LDV units, with a further 50 rounds per rifle available from command reserves. Whether or not service rifles were available the LDV took up duty within days. For a few weeks they operated as a rather loose and amorphous body – a collection of sections and platoons attached to an LDV group – but by the end of July these had been formalised into something more closely approximating to normal army structures. The LDV groups eventually became one or more battalions, each with a varying number of companies and platoons. What must always be kept in mind is that LDV units and sub-units were usually very much larger than their regular army equivalents. The 2nd City of Glasgow Battalion, for example, had 33 platoons, many of which were 150 to 200 strong – a far cry from the 30-strong platoon of the regulars. Even the 1st Dunbartonshire Battalion, which was formed with 1,586 members and was quite a small battalion by LDV standards, had about twice as many men as a regular infantry battalion of the period.

There were two sorts of LDV units created in the summer of 1940. One was what is normally thought of when the LDV or Home Guard is mentioned and was, as described in the previous chapter, a community-based home defence force. However there was also a range of units based on workplaces, whose primary responsibility was the defence of their factory, mill, mine, power station, dock or railway. Eventually these were all brought within the same over-arching structure, although not without some difficulties.

The most notable of these works units were the railway and Post Office battalions. These were most numerous in Glasgow, which eventually had two railway battalions: the 7th City of Glasgow (LMS) Battalion and the 8th City of Glasgow (2nd LNER) Battalion, and three Post Office battalions: the 9th City of Glasgow (14th GPO) Battalion, the 10th City of Glasgow (15th GPO) Battalion and 13th City of Glasgow (40th GPO) Battalion. As the unit names suggest, these railway and Post Office battalions had both a local responsibility and were integrated into local command structures as well as having a connection to their civilian employer, hence the parenthetical sequential numbering as, for example, 15th GPO Battalion. Railway and Post Office battalions usually had a much wider geographical spread than their unit title would suggest. For example the Aberdeen GPO battalion – the 6th Aberdeenshire (12th GPO) – had its headquarters and two companies based in the city of Aberdeen but had the rest of its sub-units scattered across the Highlands.

6th Aberdeen (12 GPO) Sub-Unit	Location
C Company	Inverness, Invergordon & Kyle of Lochalsh
D Company	Wick
E Company	Peterhead

Independent Platoon	Elgin
Independent Platoon	Kirkwall
Independent Platoon	Lerwick
Independent Platoon	Stornoway
Independent Section	Stonehaven

The history of the 6th Aberdeenshire Battalion explains the concept behind the formation of such units and the changes that were to take place over time in their operation: "Original intention in 1940 was to provide for the immediate defence and protection of all Post Office property, particularly where it was concerned with communications important to the defence of the country. As the Home Guard became more efficient and Regular Army static formations came into being, the Post Office Home Guard were gradually absorbed into the general defence scheme of a town or area, the immediate defence of PO property not necessarily being their duty, though in most cases it did remain so."[1]

Although these remote Post Office and railway sub-units were administered by their parent formation they were of course integrated into the local defence scheme for their geographical area. So the 2nd Scottish Borders Battalion, operating in the Galashiels, Selkirk and Melrose area, had within its operational area a detachment of the 10th City of Edinburgh (3rd LNER) Battalion in the Newton St Boswells area, which when not required for railway defence formed part of the town garrison. It also had in Galashiels a company of the 11th City of Edinburgh (11 GPO) Battalion, whose strength was drawn on to provide personnel for the battalion signals section and were also assigned a role in the defence of the town.

Both the management of the two railway companies operating in Scotland – the London Midland & Scottish (LMS), and the

London and North Eastern (LNER) – and the Post Office were particularly enthusiastic in their support of the LDV and later the Home Guard. In the case of the Post Office, Home Guard units volunteers were given a week's paid leave from their duties to attend a summer camp. These summer camps ran for several months in 1941 to 1944 and could accommodate 120 men each week. They were held at various locations including Banchory, Abernethy and Dalmahoy.

Other works units formed platoons or companies within a general service battalion; thus the 5th Fife Battalion, serving in the Kirkcaldy area, had in its Burntisland Company works platoons from the Burntisland Shipbuilding Company and the local British Aluminium plant, the latter being a Ministry of Aircraft Production site. In the larger areas a battalion would be formed from works units; in Dundee the 2nd Dundee Battalion had five companies drawn from the main local employers:

A Company	Caledon Shipbuilding
B Company	Dundee Corporation Gas & Electricity Departments & Baxter Bros., Jute manufacturers
C Company	Dundee Corporation Transport Department
D Company	Urquhart Lindsay & Robertson Orchar Ltd (Textile machinery manufacturers)
E Company	LMS & LNER

Glasgow would eventually have four works battalions in addition to its railway and Post Office units; one of these, the 6th Glasgow, recruited 2,200 men almost exclusively from among the employees of the city council with the various companies coming from different areas of the council's operations:

A Company	Transport Department
B Company	Gas Department
C Company	Water Department
D Company	City Chambers Departments
E Company	Electricity Department
F Company	Lighting and Cleansing Departments

In Glasgow more than half the total number of volunteers were members of these works units and there was a particularly urgent need for the works units to be integrated into the overall defence plan for the city as well as coping with the specific defence of their own industrial premises.

Whatever the expectations of the War Office had been for the new force the reality of the huge response required a pretty rapid degree of rethinking. On 5 June, General Sir Edmund Ironside, who was Chief of the Imperial General Staff from September 1939 to May 1940, and had succeeded General Kirke as the Commander in Chief Home Forces, called a conference at York of all LDV senior commanders. He opened by acknowledging the problems that had arisen: "I would ask you to remember that this show has been started in a very great hurry and we are working very hard to get it going. . . It does take some time, and you were formed really before we had a chance to get down to it, therefore any mistakes that have been made occurred through the pace at which we have worked."[2]

Ironside went on to review the progress that had been made: 140,000 denim uniforms and 80,000 rifles had been issued, and one million rounds of solid-shot ammunition for shotguns had been ordered, which he enthusiastically pointed out would: "kill a leopard at 200 yards". He identified the main roles of the LDV as static defence and information; the fall of

France he attributed to the lack of any system of local defence once the main defences had been breached. The advantages of a locally based force that knew its area, whether town or country, were obvious, and if they passed information swiftly to the proper authorities invaders could be dealt with before they had a chance to get established.

Ironside promised to prioritise the country areas in the supply of rifles as he had concerns about high-powered rifles being fired in city streets, an environment he thought more suited to shotguns. He also introduced the LDV to the first of many unorthodox weapons that would come its way, the "Molotov cocktail". This was a home-made device constructed from a bottle filled with petrol, and equipped with a burning fuse made from petrol-soaked cloth which, as Ironside said: "if thrown on the top of a tank will explode, and if you throw half a dozen more on it you have them cooked."[3] The Molotov cocktail, mockingly named after the Soviet Union's foreign minister Vyacheslav Molotov, had been developed in Finland for use against Soviet tanks in the Soviet-Finnish Winter War of 1939–40. Not everyone was quite so enthusiastic about the Molotov Cocktail; Tom Wintringham, the former Commander of the British Battalion in the International Brigade wrote: "They seldom can be relied on to stop a tank. If lobbed on to the top of a tank in the way that is sometimes advised, they merely warm it gently."[4]

The Commander in Chief also announced that a sum of £10,000 had been allocated to each command to pay for the small things that were needed, such as telephones. This, it is to be hoped, addressed Major Broun Lindsay's financial difficulties in East Lothian, which were discussed in Chapter 2.

Some of the administrative difficulties of the LDV were eased by the decision in June to use the county-based Territorial Army Associations as an agency to deal with administration, finance and stores. These associations had, of course, lost much of their function after September 1939, when the Territorial Army had been mobilised and incorporated into the regular army structures and systems. There had been a suggestion that the British Legion, the ex-servicemen's association, might be used in this role but the balance of military opinion favoured the TA associations. The Legion headquarters in Scotland had offered to put its resources at the disposal of the authorities in organising and recruiting the LDV when Eden's appeal was broadcast. There grew up in some areas an erroneous idea that recruitment to the LDV had to be through the British Legion; Emanuel Shinwell, the Labour MP for Seaham, County Durham, who had been a War Office junior minister in the Labour government between 1929 and 1930, asked Eden in the House of Commons on 23 July to confirm that there was no need for recruits to join through the British Legion route, an assurance Eden gave. The suspicion on the left was that the British Legion was perhaps too conservative a body to handle recruitment and that a better response and a better social mix would be obtained by using official agencies such as the police and the labour exchanges.

By the end of June, LDV area commanders were in place in the four areas into which Scotland was divided, with Colonel H.F. Grant Suttie DSO, MC, commanding the Edinburgh area, Colonel John Colville PC, MP, the Glasgow area, Colonel T.F.B. Renny-Tailyour CB, South Highland, and Colonel Usher, North Highland. These appointments of men of considerable military experience relieved the army area commanders of much of the detailed work related to LDV training and organisation.

In June Scottish Command were able to issue denim uniforms and field service caps at the rate of 10 per section. More rifles became available by drawing on Cadet Force weapons and the first deliveries of Canadian Ross rifles and, much to the delight of the

LDV, five rounds of ammunition per rifle per week were allowed
to be fired for training. As three or four men were sharing a rifle
this extremely modest allocation was hardly likely to make for a
force full of crack marksmen.

Another sign that the LDV were being taken seriously came
with the appointment of Lieutenant-General Sir Henry Pownall
as Inspector General of the new force. Pownall was 53 years of age
and had been Chief of the General Staff to General Gort, Commander in Chief of the British Expeditionary Force in 1939–40.
He was seen as a high-quality appointment, a fact which underlined the significance of the force. The Commander in Chief
Home Forces was still in charge of operations through commands
and areas but Pownall's duty was to ensure that the LDV's training and doctrine were, as far as possible, standardised.

The volunteers at a section and platoon level carried out their
duties as best they could, patrolling and watching, armed with
whatever arms they had scrounged or been allocated. At a higher
level the new body took some time to reach organisational stability, and to some extent structural changes continued throughout
the life of the force, but the lack of stability in the early weeks was,
of course, a function of the rapid and ill-considered planning that
had attended the birth to the LDV. To take one example, Fife and
Kinross was, on the formation of the LDV, designated as Group
10 of Zone IV within Highland area. This group had a notional
establishment of 1,440 volunteers organised in four companies,
with an allocation of 360 rifles and 10 rounds per rifle. The Earl
of Elgin, chairman of the Fife Territorial Army Association and
Lord-Lieutenant of the County of Fife, recommended that Fife
be organised under one commander with seven units contermin-
ous with the county's Air Raid Precautions (ARP) districts. On
10 July 1940 the Fife Local Defence Volunteers became Zone VI
of the Highland Area. However, a mere five days later a decision

was taken to split Highland Area into a North Highland and a South Highland Area, and Fife, despite hardly being a Highland county, found its third identity in two months as Zone III of South Highland Area.

These command arrangements naturally affected the day-to-day work of the LDV but they were overlaid by the arrangements for operational command that would have come into force in an emergency. For this purpose the Fife LDV came under operational orders of 27 Infantry Brigade, part of 9 (Highland) Division which was stationed in the county. Of course as regular formations moved in and out of an area, the operational control of the LDV or the Home Guard changed; thus Fife passed from being the responsibility of 27 Brigade to 138 Brigade and then, when Polish forces were moved into Fife, the Fife Home Guard battalions were under the operational direction of General Paszkiewicz.

As uniforms became available, badges of rank were formalised – although they were not actually badges of rank, as the LDV did not have ranks, but rather badges showing the appointment held. This rather esoteric distinction echoed the traditional army practice of considering lance-corporal to be not a rank but an appointment. As such, a lance-corporal could be demoted by administrative action rather than by court martial; so the concept of an appointment at specified grade was not entirely a novelty. The basic LDV unit was the section under a section commander, who wore a sergeant's three chevrons on his left arm. The section second in command wore a corporal's two chevrons and the sub-section commanders wore the single chevron of a lance-corporal. Battalion, company and platoon commanders wore, respectively, three, two and one stripes of blue braid on their shoulder straps, and of course were referred to as, for example, "platoon commander" rather than as "lieutenant". Group commanders wore four blue braid stripes and the zone commanders wore a broad blue band. The notional strengths of the various sub-units were: sections 25 men, platoons 100 men, companies 400 men. Battalions were expected to number about 1,600 men, although many were considerably stronger.

Recruits were flowing into the LDV at a satisfactory rate, arms and equipment were trickling through and the organisation on the ground was taking shape. Commanders could be identified by strips of braid and the higher command structures were evolving. However Britain was now fighting alone; Norway had surrendered on 10 June and on the same day the opportunistic fascist dictator of Italy, Benito Mussolini, declared war on Britain and France. On 22 June Britain's last European ally, France, signed an armistice with Germany. While soldiers, sailors and airmen from Norway, the Netherlands, Belgium, Poland, Czechoslovakia, Denmark and France had made their way to Britain to continue the fight, the grim fact remained that the United Kingdom and the Commonwealth were the only powers still engaged in the struggle. With the entry of Italy into the war the prospect of fighting in the Middle East became a consideration and one that became a reality in August and September. When Churchill addressed the House of Commons on 4 June at the end of the Dunkirk evacuation, his message was defiant, but it expressed the reality of what was likely to happen next: "We shall defend our island whatever the cost may be. We shall fight on the beaches, we shall fight on the landing grounds, we shall fight in the fields and in the streets, we shall fight in the hills. We shall never surrender."[5]

Churchill understood, better than most politicians of his day, the importance of morale, of public confidence, of presentational issues. He had, after all, become prime minister largely because the public and parliament had lost confidence in Chamberlain.

As First Lord of the Admiralty, Churchill had been involved in the direction of the war since its outbreak. Having been heavily involved in the decision-making connected with the failed Norwegian campaign, he was in no sense "a clean pair of hands", nor in many eyes "a safe pair of hands", but he did exude confidence and energy.

As early as October 1939 Churchill had written to a cabinet colleague suggesting the formation of a Home Guard of half a million men over 40 years of age. He eagerly responded to the formation of the LDV; the idea of a nation mobilised against the common foe, of communities banding together in the face of danger, was one that made a considerable appeal to the prime minister's romantic imagination. He could not fail to be aware of the interest being taken in the new force inside and outside parliament and asked for a report on its development. One issue he quickly seized on was the name; in Churchill's view "Local Defence Volunteers" did not strike the right note and he told Eden as much. Churchill went back to the title he had used in his memo in October 1939 and suggested to Eden that "Home Guard" would be a better and more inspiring title: "I don't think much of the name Local Defence Volunteers for your very large new force. The word 'local' is uninspiring. Mr Herbert Morrison [Labour MP and Minister of Supply] suggested to me to-day the title 'Civic Guard', but I think 'Home Guard' would be better. Don't hesitate to change on account of already having made armlets, &c., if it is thought the title 'Home Guard' would be more compulsive."[6]

Eden resisted the suggestion, partly on the severely practical grounds that a million armbands stencilled with the letters LDV had been manufactured and that the term was now part of the military lexicon, and partly on the very human grounds that the new mass movement was identified with him and a major element in his public profile. Eden was doubtless aware that if Churchill succeeded in "re-branding" the force it would inevitably become more associated with the prime minister than with the secretary of state for war. Churchill, of course, was also minister of defence, and chaired the War Cabinet, and as such had an obvious avenue to intervene in War Office matters even beyond that allowed by his role as prime minister. If Eden had feared that Churchill would increasingly intervene in the Home Guard he was proved right, and a flow of Churchillian memos on all aspects of the Home Guard would begin to emanate from 10 Downing Street

Churchill had his mind made up on the matter and used every means in his power to have the new title used. In a radio broadcast on Sunday 14 July, for example, Churchill referred to the growing strength of the British army and went on: "Behind these, as a means of destruction for parachutists, air-borne invaders, and any traitors that may be found in our midst – and I do not believe there are many, and they will get short shrift – we have more than a million of the Local Defence Volunteers, or as they are much better called, the Home Guard. These officers and men, a large proportion of whom have been through the last War, have the strongest desire to attack and come to close quarters with the enemy, wherever he may appear."[7]

This section of his speech is of interest on two counts, quite apart from Churchill's advocacy of the title Home Guard. He speaks of "officers and men" – in contrast to the official doctrine that there were no such things in the LDV, simply volunteers holding varying appointments. Churchill, as an ex-officer with military service going back to the North West Frontier in India and the Sudan Campaign of 1898, presumably thought naturally in terms of traditional rank divisions, a concept that would also be familiar and probably comfortable to the LDV members with previous military service, whether as officers or as other ranks. The prime minister also talks of the volunteers desire to "attack and come to close quarters with the enemy", and again a shift in concept can be detected. If the LDV's original role had been to watch and report then the Churchillian concept was of something much more active, and much more in keeping with his aggressive spirit and the national mood of defiance. He went on in his broadcast address to echo his "we shall fight on the beaches" speech of 4 June: "We shall defend every village, every town, and every city. The vast mass of London itself, fought street by street, could easily devour an entire hostile army, and we would rather see London laid in ruins and ashes than that it should be tamely and abjectly enslaved."

Although this street by street defence of every inch of Britain does not specifically mention the Local Defence Volunteers it is clear that Churchill envisaged them as playing a vital role in such a struggle – again a marked change from the original concept which was that the LDV would not recruit in large towns and cities and would be not much more than a rural and semi-rural special constabulary.

The official War Office position was that the name-change to Home Guard should not be made. General Pownall, the LDV Inspector General, thought that the only reason for the change was that: "Home Guard rolls better off the tongue, and makes a better headline."[8] This of course was true and while from a professional soldier's standpoint these advantages might not have been worth serious consideration it was an undeniably significant factor in terms of national morale and cohesion. The title Local Defence Volunteers had an uninspiring ring of local bureaucracy and a parish pump approach. It had not taken people long to invent humorous versions of the acronym LDV: "Look, Duck and Vanish" was perhaps the commonest but "Last Desperate Venture" or "Long Dentured Veterans" also circulated – all of which was hardly conducive to *esprit de corps* internally or public confidence in the force externally.

When Churchill's mind was made up on some strategic plan or troop deployment he could, with difficulty, be persuaded that he was wrong, but on this matter, essentially a public relations and presentational one, he was perhaps on firmer ground. His professional military advisers had no real arguments other than cost and those million LDV armbands. Churchill undoubtedly felt that he was in tune with the feelings of the British people, and he was right to do so. Contemporary studies of public opinion showed a huge degree of confidence in his leadership, despite the enormously depressing military situation.

On 23 July 1940, Anthony Eden was answering questions in the House of Commons on the Local Defence Volunteers. After dealing with a question on radios and vehicles for the LDV from Sir Thomas Moore (Ayr Burghs), ever-active in LDV issues, and a question about medals, the third question came from Seymour Cocks, the Labour MP for Broxtowe, Nottinghamshire:

> Mr Cocks asked the Secretary of State for War whether the title of "Local Defence Volunteers" is to be changed to "The Home Guard".
>
> Mr Eden: It is proposed to submit for His Majesty's approval an Order in Council, giving the Local Defence Volunteers the title "The Home Guard." Armlets will be adapted as convenient to the initials "H.G".[9]

The formal status of the LDV had been enshrined in an Order in Council (1940 No. 748) of 17 May entitled The Defence (Local Defence Volunteers) Regulations, 1940. This, as Eden indicated, would be modified by a further Order in Council (1940 No. 1383) of 31 July which added a 3rd regulation: "The Local

Defence Volunteers constituted under these Regulations may be known by the alternative title 'the Home Guard.'"

The change of title did give general satisfaction. *The Times*, in an editorial after Churchill's 14 July speech, spoke of "the new and better name"[10] and on 22 July, the day before Eden's announcement, the newspaper reported on an inspection of Local Defence Volunteers in Essex by the king under the headline "The King With Home Guard" and used the two titles interchangeably in the body of its report. The renaming of a London Midland & Scottish Railway 4-6-0 "Patriot Class" locomotive as "Home Guard" at Euston Station by General Pownall on 30 July perhaps put a seal on the success of the new name.

Chapter 5

A NATION IN ARMS

From the Rhinns of Galloway to Shetland men flocked to take up arms (had there been arms to take up) against the threat of invasion and enemy raids. A notional ceiling for numbers for the Home Guard in Scotland was drawn up by Scottish Command in August and envisaged a distribution of personnel by army areas as follows[1]:

Edinburgh Area	25,000
Glasgow Area	60,000
South Highland Area	30,000
North Highland Area	20,000
Scotland Total	135,000

These figures were never firmly adhered to and by July 1942 the breakdown of members was as follows (there had been several changes of structure and title in army structures and there is as a result no exact equivalence with the August 1940 targets)[2]:

North Highland District	26,654
West Scotland District	74,145
South Highland Area	24,710
Edinburgh Area	27,584

Orkney Garrison	1,256
Shetland Garrison	1,407
Scotland Total	155,756

The flood of recruits posed serious administrative difficulties and the rapid growth of the organisation is demonstrated by the figures for Lanarkshire (Zone II of Glasgow Area).

Date	12 June 1940	19 June 1940	3 July 1940
	Strength	Strength	Strength
Group I	1,222	1,533	2,193
Group II	2,351	2,977	9,796
Total	3,573	4,510	11,989

It is not hard to appreciate the administrative burden of enrolling close on 12,000 men in a matter of weeks, a task carried out with little clerical assistance and with, at best, improvised systems. By 24 June the Lanarkshire Zone had been organised into four battalions:

+ 1st Upper Ward (the former Group 1) under Colonel J. E. Cranstoun
+ 2nd Coatbridge/Airdrie etc under Colonel J. M. Arthur
+ 3rd Motherwell/ Wishaw/Shotts under Lieutenant-Colonel Houldsworth
+ 4th Cambuslang/Hamilton etc under Colonel Mather

By August the 4th Lanarkshire battalion was so large that the Rutherglen and Cambuslang men were split off and formed into the 5th Lanarkshire Battalion.

The War Office recognised that units of this size could not be run on a purely part-time, voluntary basis and gradually provided paid staff to undertake the essential administrative duties.

The first step on this path came with the authorisation for the employment of a paid administrative assistant for every unit with an authorised strength in excess of 1,500 men. In 1941 battalions were allowed an Adjutant/Quartermaster, a paid full-time post. Initially these Adjutant/Quartermasters had to be found from within the Home Guard's own strength but were later supplied from among the regular army's pool of officers unfit for active service. In November 1940 it was announced that a paid full-time storeman would be provided at the company level. Some staff instructors were allocated to Scottish Command and were immediately sent off on a training course to familiarise them with the non-standard Home Guard weapons; there was little point in a musketry instructor being fully proficient on the Lee Enfield if he had to teach men how to use an American Springfield rifle. In addition one or more PSIs – Permanent Sergeant Instructors – were supplied to each battalion from 1940 onwards. These last appointments were intended to improve the operational skills of the Home Guards in areas such as weapons training and field-craft, but there were some problems experienced. The sergeants appointed were usually men of low medical category, and some of them proved to be unsuited to work in a scattered rural area where training had to be decentralised to company and often to platoon level, and where there were the particular challenges of training Home Guardsmen. The history of the Dumfriesshire Zone suggests that many of these PSIs had first to be taught to drive by their Home Guard hosts before they could carry out their duties.[3]

The wave of patriotic enthusiasm that brought so many men into the LDV had, on occasion, rather less happy consequences. The Chief Constable of Glasgow, Percy Sillitoe, an officer who had won a considerable reputation for breaking up the gang culture of the city in the 1930s, experienced the ill-directed enthusiasm of the

LDV at first hand. In his memoirs he tells how the LDV saw spies, fifth columnists and invaders in any passing motorist who was not known to them and also suggests that the LDV had attracted a fair number of undesirable characters. This was bad enough but worse was to come: "...one night, after an air-raid warning, I myself was accosted by an LDV who held a loaded rifle fully cocked, pointed at my stomach on Albert Bridge. My identification papers he waved aside, and nothing that either I nor Dr Imrie, the Police Surgeon, who was with me, could say, convinced him that I should be allowed to pass."[4]

Sillitoe felt that enough was enough and, with the approval of the army area commander and the Lord Provost, inserted an advertisement in the Glasgow papers warning that the police would take action where drivers were stopped and threatened for no good reason. The LDV area commander demanded the withdrawal of the notice and told the Chief Constable to communicate with the city of Glasgow zone commander about any cases of improper stopping of traffic by the LDV. Sillitoe insisted that as long as civil law prevailed within the city he would not accept instructions from the military. Eventually Sillitoe and Lieutenant-Colonel Simpson, the LDV Zone Commander, met and reached an amicable solution.

Holding the Chief Constable of Glasgow up at gun-point was a major error of judgement but perhaps had its farcical side. Sadly not all cases of LDV enthusiasm were so trivial in their consequences. On 30 June David Calder, a quarrymaster from Leuchars, failed to halt his car when challenged at an LDV control point in Fife. A volunteer, James Lennie, had waved a red lamp but Calder failed to slow down or alter course and Lennie had to jump off the road to avoid being struck. Another volunteer, Andrew Crookston, then fired a warning shot over the car after it had passed and gave evidence to the fatal accident inquiry at Cupar Sheriff Court that he heard another volunteer, whose name does not appear in the press report, fire two shots. Calder's car swerved and crashed. Medical evidence was given which confirmed that Calder had been shot through the head and that death would have been instantaneous. The young farm worker who fired the fatal shots was, the Sheriff said: "carrying out duties in an important public service and that completely shielded him from any responsibility."[5] However, an official booklet issued to LDV members instructs sentries stopping cars as to the correct procedure to be followed: "If firing at a car for failing to stop, aim at the tyres",[6] so neither Volunteer Crookston, who fired a warning shot over the car, nor the volunteer who fired the fatal shot were acting in accordance with accepted procedures.

Inevitably accidents happened and Home Guardsmen were killed or injured in training. A particularly serious incident, on 17 August 1940, affected the 1st Perthshire Battalion. During training on hand grenades a Mills bomb exploded, killing the Company Commander, Major Frederick D. Mirrielees, and Volunteer McNiven, and injuring Section Leader A. Fraser and Volunteer R. McCallum. Major Mirrielees, of Garth House, Aberfeldy, was an experienced officer who had served in the artillery in the First World War. The Procurator Fiscal for Perthshire certified the death of Major Mirrielees as due to "internal haemorrhage probably caused by injuries by bomb splinter to blood vessel or vessels in the abdominal cavity, plus shock."[7] Volunteer Dougald McNiven, a 37-year old master mariner from Kenmore, died the next day in Perth Royal Infirmary from lacerations of the right lung and pericardium causing severe haemorrhage.

The death of Major Mirrielees sparked off a long-running legal wrangle about the status of the Home Guard. The major left an estate valued at £300,000 and the Inland Revenue claimed £64,000 in death duties. His trustees appealed against this, basing their argument on an 1894 Act giving relief from death duties to a "common soldier" who died in service. When the case came to the Court of Session in July 1942 Lord Keith ruled that it was not possible to see the Home Guard as a self-contained force comprising nothing but common soldiers, and accordingly he ruled against the trustees on Major Mirrielees' estate.[8] The matter was appealed to the House of Lords where the "equality of service" argument was found to be more persuasive than the Crown's submission that as the major had been appointed to be a Company Commander he could not be considered as a common soldier. The Lord Chancellor, presiding, concluded that as a member of the Home Guard when he was training or serving he was subject to military law and was to be considered as a common soldier. The remainder of the bench concurred and the trustees' appeal was allowed, with costs.[9]

Across Scotland the new body's character was being formed. Different areas had a different population to draw on; Aberdeenshire seemed rich in retired generals, with General Sir John Burnett-Stuart of Crichie and Major-General Sir James Burnet of Leys becoming company commanders. In contrast, other areas, such as the south-west corner of Fife where the 7th Fife Battalion recruited, were less well provided for with retired senior officers. The unit history notes that: "The area is a workaday one, with little in the way of a leisured class and containing no nucleus of ex-Regular officers, such as exists in other parts of the country, on which the early organisation of the Home Guard might have been built up. Immediately available were one or two senior Territorial officers of long service... and a certain number of veterans of other ranks, some of them ex-NCOs."[10]

The LDV was an inherently flexible organisation and functioned best when it worked with the grain of local circumstances. In the isolated and thinly populated area of Teviothead and

Roberton in Roxburghshire there were not enough men to form a normal company nor was the area easily combined with another district. Accordingly, a half company was formed, with two platoons and no company second in command. This sub-unit, G Company of the 1st Battalion Scottish Borders Home Guard, still managed to recruit about 100 men, as practically all of the available manpower volunteered.

Flexibility could even extend to winking at the participation of women in operational duties in some circumstances. The 1st Dumfriesshire Battalion, operating in the Lockerbie and Kirkconnel area, manned observation points in 1940, including one at the ancient hill fort of Burnswark south-east of Lockerbie. Here Lady Buchanan-Jardine of Castlemilk took her turn on duty along with gamekeepers and estate workers who had enlisted in the LDV. Her husband, Sir John Buchanan-Jardine, was off in the army; Lady Buchanan-Jardine was presumably running the Castlemilk Estate and clearly nobody was going to tell her she should not go on duty with the members of her estate staff. It might not have been the form of active service Dr Sumerskill would later argue for, but what happened in the depths of rural Dumfriesshire perhaps owed as much to traditional relationships and local power structures as to army council instructions.

Local circumstances were allowed to dictate structures in a way that would not have been possible in a more regularly organised force. In Angus, in the country area covered by the 2nd Angus battalion, the unit was organised on apparently traditional lines with a battalion commander, a major as second in command, and four majors as company commanders, each with four platoon commanders under them. However, in the case of the Tannadice Platoon in B Company the platoon commander was also the battalion second in command – a pragmatic response to the presence of an experienced officer in the Tannadice area. Due to the local

and devolved structure and role of the Home Guard it would be
unusual for the whole 2nd Angus battalion to be deployed as a
unit and this meant that Major Neish, the battalion second in
command, could devote most of his attention to the command of
the substantial body of men in the Tannadice Platoon – around
200 in all ranks. The command of the Tannadice Platoon and
the overlap of duties at a higher level was facilitated by its sub-
division into a platoon headquarters under an admin. officer of
2nd lieutenant's rank, with 15 other ranks as clerks, telephonists,
etc. and five operational sub-platoons. Three of these were static
platoons charged with operating roadblocks; in one case, the Upper Glenogil Static Platoon, a sergeant and 20 men, were tasked
with guarding two waterworks and a telecommunications site.
The other two platoons, the Finavon Mobile Battle Platoon with
one officer and 37 men, and the Justinhaugh Mobile Battle Platoon with one officer and 39 men, were specifically tasked with
dealing with air-landings and the support of the key defensive
points manned by the Finavon and Glenogil Static Platoons. The
Tannadice Platoon should thus not be confused with a platoon in
the regular army which would have comprised one junior officer
and about 30 other ranks; the Tannadice Platoon had a major,
5 lieutenants, 6 sergeants and 31 corporals and lance-corporals
and 154 privates, and was thus substantially larger than a regular
army company.[11]

The Home Guard brought all parts of the community into
contact with each other in ways that would not otherwise have
been possible. In the ranks of the 3rd Edinburgh, which had its
battalion headquarters in the Braid Hills Golf Club, served David
Pinkerton Fleming, who in civil life was a Senator of the College
of Justice – that is, a judge in the Court of Session and the High
Court of Justiciary. Lord Fleming found himself on patrol one
night with a fellow volunteer whom he had last seen in the dock

before him in the High Court. Lord Fleming was later commissioned in the 3rd Battalion.

The Home Guard was less socially divided than the regular army had traditionally been and there was a significant amount of movement between "other ranks" and commissioned status. Officers tended to be drawn from a wider range of backgrounds than had normally been the case in the regular army. Number 7 Company 4th Scottish Borders Battalion perhaps exemplifies some of these features. The Company Commander, Major J.G.G. Leadbetter, by profession a Writer to the Signet, might be taken to represent the more traditional "officer class"; he had served in Gallipoli and Italy, been attached to the Imperial Camel Corps in Palestine in the First World War and was a member of the Royal Company of Archers, the King's Bodyguard in Scotland. Major Leadbetter's four platoon commanders, however, included two farmers, a chauffeur and an estate lorryman, a somewhat wider cross-section of civilian occupations than might have been expected in an officer's mess in the regular army. The same company's records[12] show the Home Guard career path of John Briggs, a gamekeeper from Edgerston, near Jedburgh. Promoted to corporal in March 1941, he was made sergeant in October of that year, and was commissioned as a 2nd lieutenant in July 1942, being finally promoted to lieutenant in January 1944.

One noteworthy feature of the voluntary spirit of the Home Guard was that it was far from unusual for members to drop rank at their own request if for any reason they found themselves unable to devote sufficient time to the force or were not able to operate at a satisfactory level. A sergeant of 7 Company, 4th Scottish Borders Battalion thus voluntarily reverted to the rank of private, and the records of the 1st Glasgow Battalion show the Battalion CO, faced with work pressures, stepping down from command and reverting to the rank of major and taking over as second in command.[13] A more dramatic voluntary demotion is recorded in the history of the 8th City of Edinburgh Battalion. This was formed from a disparate collection of units, including the dental and veterinary students from the University. Lieutenant-Colonel Robertson was appointed as the first commanding officer of the unit, but before it became operational he was obliged by illness to resign his command and reverted, at his own request, to the rank of private.[14]

So enthusiastic was the response to the Home Guard that in July Eden was obliged to announce a temporary suspension of recruitment, other than in areas where the quota had not yet been reached, this suspension to allow time for equipment and organisation to catch up with the flood of men. At this stage over 1.3 million men had signed up across the UK. One of the areas where more men were still needed was Edinburgh, and Colonel Blair, the Area Organiser, told the *Scotsman* that "owing to increasing duties in that city there were increasing openings for volunteers."[15] One of these tasks was carried out by the 1st Edinburgh, covering the western outskirts of the city, which spent the summer months of 1940 filling thousands of sandbags with the fine sand from Cramond Beach and transporting the filled bags to create strong points and defence posts across the battalion area.

One of the early problems faced by the Local Defence Volunteers was that many of the best, fittest and most public-spirited men in a community had already signed up for one of the other Civil Defence services and were working in Air Raid Precautions (ARP) or the Auxiliary Fire Service or in the Police War Reserve. Eden's call had said that volunteers in these services could join the LDV with the approval of their superior officers. Towns and cities which were likely to be bombed would obviously need to maintain a very high level of civil defence preparedness and would have difficulties releasing men for the LDV.

On the other hand many rural volunteers in ARP and other services, who had joined before the LDV was conceived and realised that their village was unlikely to be bombed but might well be on the invasion route, saw the new force as a more effective way of "doing their bit" for the national struggle. In Bearsden 42 ARP wardens enrolled in the 3rd Dunbartonshire Battalion in order to gain the legal right to bear arms. The Officer Training Corps units at the four Scottish universities also volunteered for the LDV, as their existing position, rather strangely, did not give them combatant status.

The competing priorities of the various services could lead to friction, but in many cases the LDV and the ARP worked harmoniously together and in Fife, where rural wardens were allowed to enlist in the LDV but wardens in towns were not, the two services worked very closely together, the LDV/Home Guard zone headquarters initially sharing premises with the ARP in the County Buildings in Cupar. ARP communication facilities were used for the Home Guard and when the Home Guard became better organised their services such as dispatch-riders, signals sections and carrier pigeons were used for ARP purposes.

In the Fife mining area of Cowdenbeath and Lochgelly the 6th Fife Battalion found that although many of the best men had already been signed up for civil defence and would not be released, the officials in the local collieries had been excluded from these services and the battalion was able to recruit many of its leaders and eventual officers from this source. Ninety per cent of the 6th Fife's manpower came from the mining industry and the strength and physical fitness of these recruits could be taken for granted, with obvious advantages for field training. The men moved fast on exercises and thought nothing of wading waist-deep in streams and bogs. One 6th Fife man was asked about this by a visiting staff officer and commented: "I work for eight hours a day in worse than that sir!" One other advantage of drawing the unit's strength from the pits was that the shift system meant that it was possible to have almost 24-hour-a-day manning. Of course the physical strain on men who had done an eight-hour shift underground and then reported for Home Guard duties was not to be underestimated, and as we shall see, as the war went on, concerns were expressed about this.

Miners and explosives proved to be a valuable asset in Dumfriesshire where in the early days of the LDV the Kirkconnel Company manufactured their own home-made grenades, thanks to the generosity of the local Fauldhead Colliery. Empty tins were filled with Samsonite explosive and ½-inch diameter punchings from steel plate. These improvised grenades weighed from one to three pounds and were fitted with a 5- to 7-second fuse and thrown by hand. Land-mines of about 7lb weight were improvised from explosives tins used in the pit and set off using an electric mining detonator. Five hundred grenades and mines were made in this way before "official" supplies became available. The Dumfriesshire initiative would have been applauded by Tom Wintringham, who, writing with his International Brigade experience, advocated the construction of large quantities of home-made grenades using the plentiful stocks of explosives that were commercially available. Large hand-grenades, containing a pound or a pound and a half of explosive were, he said, in the hands of brave men "far more efficient than either anti-tank guns, petrol bombs or tank-traps".[16]

The 1st Wigtownshire Battalion created their own form of Molotov cocktail even before this was formally recognised as a Home Guard weapon. Lord Stair, the Wigtownshire Zone Commander, bought up all the safety matches in the county to ensure a supply of igniters for these devices. A considerable amount of local initiative in the construction of ingenious means of bringing death and destruction to the enemy went on across Scotland. The 4th Perthshire area was particularly inventive in this respect; a Mr J. Cameron, of Bullionfield paper works at Invergowrie, invented a new form of fuse for the Molotov cocktail. These were supplied to other battalions, and at one stage one of the battalion's officers had five women employees in his factory making them. The same factory also produced a carriage to allow a Home Guard flamethrower to be towed behind a car or lorry. However, not every development was successful; in the same battalion area a form of phosphorous bomb was invented at the Lumsden & MacKenzie bleach works in Perth, but rejected on the grounds that it produced dangerous quantities of phosphorous pentoxide. In West Lothian the first batch of Molotov cocktails was made by Scottish Oils Ltd., the distillers and refiners of the local shale oil.

The LDV were to be issued with denim overalls as opposed to normal army battledress; by 22 May the under secretary of state was able to tell the House of Commons that 90,000 suits of overalls had been issued and that a quarter of a million field service caps were available. A similar number of armbands stencilled with the letters LDV were on order. In reply to a question from Mr Henderson Stewart (Liberal, Fife East), the Under Secretary Sir Edward Grigg reiterated that field service caps, and not steel helmets, would be issued. The overalls and headgear issue came to have considerable significance and a change of policy was brought about, largely due to pressure from the force.

The decisions not to have ranks or standard insignia of office, to issue overalls rather than battledress, and caps rather than helmets, were taken for a variety of reasons. There was a shortage of battledress uniforms and steel helmets, but also very clearly a desire to avoid making a voluntary force seem too close in style to the regular army, partly due to the initial very limited conception of the role of the LDV; if it was simply an armed special constabulary, did it need steel helmets? One of the rather strange consequences of the equality of ranks policy, which owed more to a desire to ensure the subordination of the LDV to the regular forces than to any hitherto unsuspected socialist and egalitarian tendencies manifesting themselves in the inner reaches of the War Office, was that a major-general (retired) acting as a company commander of the LDV was, technically, a private soldier and thus outranked by every officer and NCO in the army.

Some concern was raised by the practice of retired senior officers who had enlisted in the LDV coming on parade in the uniform of their former rank, and Anthony Eden announced in parliament on 13 August 1940 that orders were to be issued banning this practice.

The issue of field service caps caused particular concern in Scotland. There was a strong feeling that the Balmoral, which the army referred to as the Bonnet ('Tam o' Shanter), was a more appropriate item of dress for Scottish troops. As early as September 1940 suggestions were going to the War Office that either the Glengarry or the Bonnet (TOS) would be preferable to the field service cap. In November the Earl of Fife, writing as President of the Fife Territorial Army Association and with the support of many other lord lieutenants and lord provosts, conveyed to the War Office the opinions of the Scottish TA Associations, which were 26 to 2 in favour of the Bonnet (although Nairn, with severe practicality, pressed for steel helmets.) The Director General of the Home Guard minuted, with just a detectable trace of weariness:"History shows that sooner or later we generally have to defer to Scottish feeling in the matter of dress. . ."[17] and proposed a compromise. Units could choose either field service caps or the Bonnet so long as there was uniformity in the battalion and the units met the difference in cost between the two items of headgear.

The Home Guard may not have been well-equipped but even in its early stages of development it was able to carry out a growing range of useful functions. In the summer, autumn and winter of 1940 the 3rd Lanarkshire Battalion maintained sunset to sunrise observation posts at Overtown, Newmains Bing and on the roof of the Shotts Cinema. In December 1940 the 7th City of Glasgow Battalion – an LMS Railway battalion – obtained the use of a goods van from the LMS Polmadie depot, armoured it with boiler plating and sandbags, mounted a Browning machine-gun on board and patrolled the line to Rutherglen.

Across Scotland vulnerable points such as power stations, bridges and gasworks had night guards provided by the local Home Guard unit. The 6th Glasgow carried out day and night patrols on the city's main water supply pipeline from Loch Katrine. In Renfrewshire the 3rd Renfrewshire had a defence scheme in place by 3 June and manned roadblocks between dusk and dawn using ex-servicemen who had formed a large element in the battalion's initial recruitment. Weapon shortages prevented this battalion from manning other defended localities but a variety of observation posts were established.

Unscheduled happenings were also taken care of by the Home Guard. On 16 September 1940 a German Heinkel 115 seaplane came down at Windyheads Hill, Aberdeenshire and the three crew members were taken prisoner by the men of the New Aberdour Platoon of 1st Aberdeenshire Battalion. What the unit history claims as possibly the first example of the Home Guard firing on the enemy came in the summer of 1940 when a German aircraft carried out a low-level bombing raid on the British Aluminium factory at Burntisland. Men of the 5th Fife Battalion fired with automatic weapons from the factory roof at the raider and claimed hits. The fact that the factory unit had automatic weapons at this early date probably reflects its status as a Ministry of Aircraft Production site and the more generous scale of provision that this status secured.

The Home Guard took its duties seriously. In August 1940 the 2nd Dunbartonshire Battalion sought permission to practise controlling the three swing bridges on the Forth and Clyde Canal in their area and asked the LMS Railway's commercial manager at Central Station in Glasgow to allow them access to the bridges from 5 to 8 p.m. on 9 August. A year later a report from the 2nd Dunbartonshire shows that each of the three canal bridges in their operational area had a team of six men trained in the operation of the bridge; a named officer was responsible for opening or closing the bridge in the event of an emergency and the bridges were incorporated into local defence plans. The Kilbowie Road Bridge, for example, was the responsibility of C Company, raised from the Singer Sewing Machine Company, Clydebank, and would be covered by riflemen and flame throwers.

With the benefit of hindsight a successful German invasion in the autumn of 1940 seems improbable, but it cannot be over-emphasised that in 1940 matters had a very different complexion. The chiefs of staff were receiving alarming news about German preparations: a concentration of transport ships in the Low Countries and France, a build-up of bombers and fighter-bombers in airfields along the Channel coast. These indications, together with favourable moon and tide states, led the chiefs of staff on the evening of Saturday 7 September to issue the codeword "Cromwell", the warning of an imminent invasion. This called Home Forces, including the Home Guard, to a state of immediate readiness. In the Highlands, the 1st Ross-shire Battalion, whose area stretched from sea to sea and across the Minch to Lewis, got the message and had circulated it to all posts within half an hour, and within two hours all posts were manned. The "stand-down" order was received at mid-day on Sunday the 8th.

One of the problems revealed by the "Cromwell" stand-to was the difficulty of communicating orders and information over large areas, and in a period when telephones were far from common. Most private houses in 1940 did not have telephones and while police stations and ARP posts might have sets, the GPO network was far from robust; and of course this was still the period of operator connection rather than subscriber connection. Although initially little official emphasis was placed on signals within the Home Guard the battalions usually developed their own communications systems, regardless of official doctrine. The 2nd Fife's unit history noted: "Although the formation of a signals section was not encouraged by Higher Authority, it was appreciated by this Unit that without a sound system of rapid communications throughout the Battalion area, any form of attack or defence was doomed to fail."[18]

The existing communications systems – GPO network, railway telephone systems and runners – were developed in the 2nd Fife area (St Andrews and environs) by local initiative: "Aldis Lamps, Buzzers, Field Telephones and Cable, Morse Flags etc were purchased and/or fabricated without expense to the country, and important Observation Posts and Vital Points were linked up by these means. Moreover, a number of volunteers owning motor cycles were formed into a small Despatch Rider Section."

In the southern outskirts of Glasgow the 3rd Renfrewshire Battalion had, by August 1940, created a field telephone system linking their companies in Clarkston, Giffnock, Whitecraigs, Barrhead and Neilston, and all the company outposts, to the battalion headquarters, a network which involved 60 miles of cable being laid. It had been found that the GPO system got seriously overloaded during incidents and exercises and this dedicated field telephone system freed the battalion from reliance on the GPO network and, of course, made capacity on the network available for other emergency users. Later this system would be supplemented by a limited number of wireless sets.

A similar development took place in the Stewarton, Troon and Irvine area covered by the 3rd Ayrshire Battalion. Their early exercises had demonstrated the fragility of the area's GPO system when subjected to heavy traffic, and the battalion's signallers laid 40 miles of cable linking battalion headquarters to all the company command posts as well as many platoon HQs and observation points.

These locally driven developments of signals services underline the considerable autonomy that Home Guard battalions enjoyed. There may have been no initial official support for a strong signals section but if battalions raised the cash and did the work the Higher Command had little alternative but to accept the position with as much grace as it could muster.

A major in the 3rd Perthshire Battalion, R.J.B. Sellar, in civilian life an author and playwright, wrote a series of articles about the Home Guard for *The Scots Magazine*. In one of these he underlined the role that local fundraising and enthusiasm had to play: "The force has been described, accurately, one believes, as 'the cheapest army in the world.' Units discover that they have to organise dances and whist drives in order to pay for materials, instruments and training gadgets which they feel ought to be issued to them. Government grants seem to be inadequate. The officers find themselves having to dip with fair frequency into their wallets."[19]

In August and September 1940 some gradual shifts of policy could be detected. A decision was taken to allow Home Guard battalions to wear the badge of their local regular infantry regiments. This was a popular move and one which made the Home Guard feel part of the local regimental family, especially as many of the older guardsmen would have served with their local regiment in the First World War. However, this decision, or rather the implementation of it, caused some anguish in Edinburgh. The initial intention had been that the Edinburgh battalions should wear the badge of the Royal Scots and many thousands of Edinburgh volunteers bought Royal Scots badges for their caps. However, a change of official mind took place and it was ordained that the Edinburgh battalions should be associated with the King's Own Scottish Borderers, on the grounds that a subsidiary title of the KOSB was "the City of Edinburgh Regiment". As an angry letter-writer to the *Scotsman* observed, after wondering if there was ever going to be any continuity in Home Guard policy: "The KOSB is a Border regiment, which recruits and has its depot in the Borders ... The Royal Scots, on the other hand, is essentially an Edinburgh regiment, which does most of its recruiting in the city and district, and has its depot in a nearby village."[20] Later the policy was changed and the Royal Scots badge reinstated.

If a regimental cap badge helped the Home Guard feel like "real soldiers" the process was greatly assisted by the decision to issue khaki battledress instead of the unpopular and shapeless denim overalls. Churchill in a speech had said that the Home Guard was as much a part of the army as the Grenadier Guards; gradually the image was matching the rhetoric.

As originally devised the LDV had volunteers (who equated to privates) and section commanders with three chevrons on their left sleeve (who equated to sergeants). Very soon, however, it was felt necessary to create other grades. Despite strong feelings from within the Home Guard and indeed from General Pownall, the Inspector General, that these appointment holders should be described as sergeants and corporals, this was resisted at the highest level. On 10 August 1940 a meeting of the Secretary of State for War, the Chief of the Imperial General Staff, the Adjutant General, the Permanent Under Secretary of State at the War Office, the Commander in Chief Home Forces and the Inspector General, Home Guard determined that there should only be two promoted appointments below Platoon Commander – namely Section Commander (three chevrons) and Squad Commander (two chevrons) and the terms sergeant and corporal would not be used. Despite this high-level decision the pressure to create additional appointments and ranks continued. On 23 August the Director General Welfare and Territorial Army, with the support of the Inspector General Home Guard argued for a patrol-leader post (analogous to a lance-corporal) to be created. A tetchy minute from Sir Frederick Bovenschen, Deputy Under Secretary of State at the War Office – a very senior civil servant – followed: ". . . this is another instance of how impossible it is to get any instruction connected with the Home Guard to last for more than a few days. I can quite understand that a growing organisation continually outgrows its suits, but this particular baby seems not to be able to wear the same clothes for more than a week at a time." By the end of September, despite his earlier position, Anthony Eden, the Secretary of State for War, was minuting his approval for the appointment of patrol leaders to wear one chevron.[21]

In September the army council set up a sub-committee on the Home Guard under Sir Edward Grigg, the Under Secretary of State, and when this reported in October it recommended the greater integration of the Home Guard into the army administrative training and command structures. Despite repeated statements by Eden and Grigg about the merits of single-status and the absence of ranks within the Home Guard, this too changed. As late as 19 October 1940 Grigg was quoted as stating that he did not like the suggestion that the Home Guard needed to be reorganised on disciplinary lines. However, following the army council decision he rose in the House on 6 November and performed an agile *volte face*, claiming to have recognised some of the problems caused by the absence of ranks. Not least of these was the fact that a Home Guard officer might be the most experienced and suitable commander of troops in an area during an invasion but could not be placed in command of any but Home Guard troops and would be outranked by any regular officer or non-commissioned officer who happened to be on the spot. He spoke of the government's anxiety to maintain the perceived advantages of the present informality of the Home Guard and to avoid introducing an undesirable rigidity into the relationships within the force. He went on: "We are, indeed, satisfied that the necessary change can be made without consequences of that kind. His Majesty has therefore been pleased to direct that King's Commissions shall be granted to all approved commanders in the Home Guard, and that the Force shall also have a suitable complement of warrant and non-commissioned ranks. The commissioned, warrant and non-commissioned officers will bear the traditional titles of their ranks."[22]

This militarisation of what had to some extent been a semi-civilian force was met with a mixed reaction. While generally it was welcome there were some on the left who saw it as the end to their dreams of a people's anti-fascist militia which would not only defend the nation against foreign invaders but defend democracy against domestic appeasers and potential Quislings. George Orwell wrote in December after these changes: "the Blimp mentality has made a big come-back."[23] His reference was to the character "Colonel Blimp" created by David Low in a cartoon series in the *Daily Express* in the 1930s. Colonel Blimp was the archetypal pompous, choleric, reactionary retired senior officer that Orwell, Wintringham and many on the left felt had too much influence in the senior ranks of the Home Guard and the upper echelons of government.

Commissioned officers in the Home Guard would enjoy sen-
iority after regular army commissioned officers of the same rank
but above regular officers of lower rank. It was envisaged that
they would not generally exercise command over regular forces
but their new commissioned status allowed the sensible utilisation of the most appropriate and experienced personnel in an
emergency.

As practical as these measures were, there was also a strong
psychological and emotional reason for the change; the need to
keep the Home Guard happy was a significant factor in policy-
making. As Grigg said: "It is, indeed, natural to assume that a
Force which forms part of the Army should be commanded by
colonels, captains and corporals, as the rest of the Army is; and
we have strong evidence from all ranks of the Home Guard that
it will never feel assured of full and unassailable military status
while these familiar ranks and titles are denied."

So, with little regret, the thin blue shoulder stripe and unfa-
miliar title of platoon commander disappeared, to be replaced by
pips and the rank of lieutenant and all the other familiar insignia
and terminology of commissioned, warrant and non-commissioned officers. However, with a truly British sense of compromise the new Home Guard commissioned officers were still for
all other purposes considered as private soldiers. If wounded or
killed on active service or training they would be pensioned as
private soldiers; if admitted to a military hospital it would be to
"other ranks" accommodation and not to an officers' ward; and
if travelling by train on official business they would be paid for
third-class accommodation rather than the first-class travel an officer was entitled to. Commissioning did not give the new Home
Guard officers any disciplinary powers; the new officers still had
to lead by example and catch the interest and imagination of their
men, men who were still giving up many hours a week to the

force, without pay, and who could resign from the Home Guard on a fortnight's notice.

Arrangements for commissioning the new officers were outlined by Sir Edward Grigg two weeks later. Each command would set up one or more selection boards to be chaired by a retired officer with Home Guard experience and would include the assistant military secretary of the command, the only regular officer on the board and the Home Guard liaison officer. The board would first of all recommend area and zone commanders and when they had been appointed they would be added to the panel for the selection of battalion commanders. When the latter were appointed they would be involved in the selection of company commanders, who in turn would be consulted on platoon commander appointments.[24]

More detail was later given on the qualities to be sought for in potential Home Guard officers:

> Officers will be chosen primarily for their powers of leadership and the confidence which they are likely to inspire in all ranks. In this respect, Boards are warned particularly, that business, social or political prominence will not be regarded as conferring by themselves powers of leadership
>
> In the case of factory, railway, Post Office and other special Home Guard units, the Board should take particular pains to ensure that the persons recommended for appointments are acceptable both to the employers and employees, and should be prepared to call in the establishment officers of the various Ministries concerned for their advice.
>
> Due note will be taken in considering candidates, of the degree of efficiency which has already been attained by the Home Guard units which they are already commanding.
>
> No candidate will be considered who is not already

serving in the Home Guard, nor will any candidate be considered who refused military service in the last war.

Interviews with a candidate in person will not be necessary where his suitability and military qualifications are
already known.[25]

On the pro-forma to be completed by applicants for commissions, two questions were asked of those who had not previously
served in the armed forces: whether they were a British subject by
birth or naturalisation, and, rather more contentiously, whether
they were of pure European descent. The implication of the latter
question was that non-white officer candidates would be unlikely
to succeed in their application.

This fairly radical set of principles would, in theory, make the
Home Guard officer corps an entirely meritocratic, if racially uniform, body. As we shall see, the practice was not entirely in line
with the precept.

Grigg's statement also recognised the size and significance of the
Home Guard, which he said was now five or six times as large as
the peacetime Territorial Army. There had already been appointed
an Inspector-General of the Home Guard, responsible for training,
but in addition it would now have a Director General. Lieutenant-
General Eastwood was appointed to this post. Deliberately chosen
as a leading young general who was lately in command of a regular
division, his appointment was intended to signal the army's commitment to the Home Guard. Eastwood would be succeeded by
another relatively young officer, Major-General Bridgeman (the
2nd Viscount Bridgeman) in 1941. Alterations were also made to
the administrative arrangements for the Home Guard, and the role
of the County Territorial Army Associations was strengthened. Indeed all TAA secretaries who were employed in other roles were
ordered to return to their associations to supervise this work. At

Scottish Command Lieutenant-Colonel Wynne, the GSOI staff officer responsible for Home Guard matters, returned to the Edinburgh TAA and was replaced by Lieutenant-Colonel Hutchison, formerly of the Lanarkshire Yeomanry.

On the island of Barra the novelist Compton MacKenzie was appointed to command the local platoon of the Outer Isles Company of the Inverness Battalion. Some of the problems and absurdities connected with the Home Guard's activities found their way into his 1943 novel *Keep the Home Guard Turning*, set on the fictional islands of Great and Little Todday; many of the characters, including the Home Guard commander Captain Waggett, also appear in his rather better known 1947 novel *Whisky Galore*. Interestingly enough the army soon decided that the ancient division of the Outer Hebrides between Ross-shire (Lewis) and Inverness-shire (Harris and the islands to the south) was militarily nonsensical and the Hebrides Battalion was eventually formed in April 1943. However it took until 1975 for local government boundaries to be redrawn to create a single civil government authority for the Western Isles. The mismatch between the requirements of military life, as interpreted by MacKenzie's "white settler" Captain Waggett, and the life-style of the Hebrides provides much fertile ground for comedy, not least in the depiction of military exercises breaching the Sabbath of the Protestant Great Todday.

As usual, however, real life imitated art. The Church of Scotland Presbytery of Deer, meeting at Maud, Aberdeenshire in September 1940 received a petition from the Kirk Session of Rathen West Church, near Fraserburgh, drawing attention to the problems being caused by the training of Home Guardsmen during the hours of religious worship. The Clerk to the Presbytery, the Rev. Dr M. Welsh Neilson, Minister of New Deer South, and himself a member of the Home Guard who had served as a chaplain to the forces in the First World War, was quoted in the *Scotsman* as saying: ". . .that, without giving away any military secret, it could be said that the training of the Home Guard was a case of extreme necessity to which there was no alternative. He pointed out, however, that later on there might be a possibility of the present stringent training being relaxed, and, in that event, the powers that be should consider leaving men as free as possible to attend divine worship at traditional hours."[26]

The conflict between Home Guard training and church attendance was to become a matter of some significance in later years, particularly as the danger of invasion was seen to recede, but Rathen West's Kirk Session was rather unusual in raising objections at a time when the invasion threat was active and real.

The Rev. Dr Neilson was not unusual in being a fighting clergyman. The 3rd Stirlingshire was originally commanded by Lieutenant-Colonel John J.S. Thomson, MC, the minister of Larbert Parish Church, who had served in the ranks and as an officer in the First World War. Across the country ministers like the Rev. J.F. McCreath, commanding the Mertoun Platoon in C Company 2nd Scottish Borders Battalion, combined the care of their parish with the defence of their country. In rural Angus, the parish minister of Oathlaw had raised a local section of the LDV but later resigned his command. He features in the Tannadice Platoon records as Private Rev. A.J. Oliphant in charge of the platoon's first aid post. His manse of Oathlaw was, at action stations, the first aid post, the pay office and the rations centre.[27]

Sunday parades and exercises were just part of the very considerable time commitment implicit in membership of the Home Guard. The 3rd Scottish Border Battalion, based in Berwickshire, was perhaps typical of a rural unit and devolved training down to platoon level. Each platoon paraded three Sundays out of four and twice nightly during the week. This level of commitment continued until 1944 when the twice-weekly parades were reduced to one parade.

In the Carse of Gowrie the 4th Perthshire Battalion had about nine parades a month for rank and file, in addition to special lectures and exercises. However officers usually found that they had to devote four or five evenings a week as well as Sundays to their Home Guard duties and most companies held an extra parade a week for officers and non-commissioned officers. Specialist courses were held over a number of Sundays for snipers, wireless operators and for training officers and section commanders.

Of course as a voluntary, unpaid, part-time service it was impossible to expect 100 per cent attendance at parades and training. Training schemes and staffing arrangements had to be devised to take account of this and specialist posts were in many units duplicated to ensure coverage. Despite the obvious problems most units reported a very high level of attendance and enthusiasm, although there were some grumbles about the quality of training; an anonymous letter-writer to the *Glasgow Herald* signing himself "Home Guard" complained that, "one has too frequently left parades with the feeling 'another night wasted, nothing done.'"[28] Another Home Guard writer to the same paper, signing himself with the traditional soldier's identity "Thomas Atkins", however ,emphasised the keenness and enthusiasm of all, especially the platoon and section commanders, in getting the organisation off the ground in four months. A considerable correspondence enlivened the columns of the *Herald* reflecting the very special ethos of the Home Guard movement and its members' willingness to question policy and argue about the direction in which the force was developing. One correspondent, who disdained the use of a pseudonym, James McWhirter of Barrhead, argued that the training regime was unsuitable; the Home Guard was being trained for the last war, not this war, and younger men needed to take on the training role.[29]

Mr McWhirter's point had some force but in practical terms there was probably very little option in the short term but to use the veterans of the First World War to train the younger men who had no military experience. The 3rd Lanarkshire Battalion, recruiting largely from miners and munitions workers in reserved occupations, found no alternative to the use of older men as trainers: "The training of these vast numbers of men was, in the first year, entirely in the hands of Ex-Servicemen from the 1914–18 war, who gave up every spare day and spare hour to train the volunteers. Without the assistance of these Ex-Servicemen, the training of the LDV and/or the HG would have been impossible. Not only in training but in Admin in the 'Q' side of the companies, it was these Ex-Servicemen and old soldiers who were the key men of their Companies. They of course received no payment of any sort for their services."[30]

Not all commanders were so enthusiastic about the value of First World War veterans. The writer of the 1st West Lothian account felt that they were ineffective except for drill and musketry training, which they conducted by rote, and that it proved to be difficult to displace these non-commissioned officers even when in time younger men without previous military experience became available and were found to be more efficient. The same writer highlighted another problem which reflects the strong local loyalties (or pernicious parochialism) of many small Scottish communities. "At no time, especially in the early days, was it found easy to employ the more talented Home Guard officers or NCOs as Instructors away from their immediate locality. What the men would learn from their own immediate leaders or from regular NCOs (when available), they would not willingly learn from someone from the next village."[31]

There was a real shortage of trainers; the rapidly expanding regular army had a big enough task to train its host of new recruits without taking on the task of catering for the Home Guard. The Under Secretary of State, Sir Edward Grigg, had expressed confidence in the capacity for self-training in the Home Guard: "The Home Guard is simply full of talent available for training as instructors, and I am certain that it will be able to find its own instructional staff."[32]

One of the distinctive features of the Home Guard was that as a voluntary body it had to be commanded with a certain degree of sensitivity and imagination and the enthusiasm of the men captured and harnessed. A comment on training in the use of the Molotov cocktail from 3rd Stirlingshire underlines this: "An experienced regular army officer from Stirling Castle gave a demonstration in the use of Molotov Cocktails, which so impressed the members of the LDV that they were now confident they had something which could be effectively used against enemy tanks. This demonstration brought home to the Officers the advisability of demonstration before the issue of any new weapons, and was acted on, on all future occasions."[33]

The Home Guard had been formed in May and had spent the summer and autumn months in a frenzy of training and preparation to meet the threat of invasion or sabotage. With the onset of winter and shorter daylight hours the opportunities for outdoor training were restricted and attention had to be paid to a winter programme. In December the Ministry of Information felt that it had to set the record straight regarding the Home Guard's role. It emphasised that the role of the Home Guard remained unchanged but with the shorter days lectures and training in buildings, lent or hired, would take the place of weekday evening outdoor parades. Higher command emphasised that local Home Guard officers should arrange training and duties so that they interfered as little as possible with members' ordinary occupations and that guardsmen should not be pressed to resign because they were unable to spare much time for duties. There had been concerns, particularly among agricultural workers, that their Home Guard duties might be structured in such a way as to prevent them carrying out their normal work. There were also ill-founded concerns that the Home Guard might be called out for full-time training; these were authoritatively denied by Eden and Grigg at the end of November and the part-time voluntary nature of the Home Guard was emphasised, along with the guarantee that mobilisation would not be considered unless there was an actual invasion.

If there was some doubt about the winter programme for the Home Guard there was a large body of experts offering to provide guidance. The flood of commercially produced training materials and booklets, mostly in compact form to fit a battledress pocket, was supplemented by Home Guard columns in daily and Sunday newspapers and in popular magazines. The journalist and Home Guardsman John Brophy was one such writer with a regular column in the *Sunday Graphic*. His *Home Guard, a Handbook for the LDV* had gone through nine reprints by October 1941 and continued to use the terms Local Defence Volunteer and Home Guard interchangeably in the body of the text. Brophy outlined his vision of the Home Guard's role: "...the first duty of the Home Guard, in point of both time and importance is: Observation and Reporting".[34] However this was tempered by his argument that: "The place of the LDV is not behind the Navy, the Army and the Air Force, but at their side. Indeed, if the enemy makes landings by parachute, troop-carrying planes or sail-planes (gliders), or if he raids inland, the LDV in many areas will probably be the first troops to resist him. And secondly, although the job is defensive, defence often means taking the attack." He also recognised that a conflict in which the Home Guard would be involved would not be a repeat of the static trench warfare of the First World War, which was the sole military experience of most Home Guardsmen, but that something much more complicated was likely: "War begins to look like a medley of small conflicts, never static for a moment, made three dimensional by air power, with no guarantee that any area involved is free from enemy troops or the civilian population." Brophy's writings, and those of the other authors of these extremely popular books, tended to reinforce the Home Guard's growing confidence in itself and its appreciation of its own value and contribution.

In fact these publications were rather frowned upon by the War Office, which had issued Home Guard Instruction No. 20 in November, emphasising that only official publications should be taken as guides to training. This was in part due to the problems caused by the different emphases of the various freelance writers, but also because of the critical, anti-establishment tone of some of the publications. However, as with many instructions relating to the Home Guard, there was a degree of disconnection between the orders handed down from the War Office and what actually happened on the ground. John Langdon-Davies's *The Home Guard Fieldcraft Manual* presents an interesting case in point. Although Langdon-Davies was Commandant of the South Eastern Command Fieldcraft School at Burwash, Sussex, his little book was a private publication which first appeared in February 1942, followed by a second edition in April 1942, so there is little evidence that such unofficial publications had dried up in the face of War Office disapproval. More contentious writers, such as Hugh Slater, former Chief of Operations of the International Brigade in Spain, produced, and one must imagine sold, works like *Home Guard for Victory! An Essay on Strategy, Tactics and Training* published by the left-wing publishing house of Victor Gollancz in

1941. There appeared to be a strong and continuing appetite for unofficial instructional books by Home Guardsmen of all ranks, despite official disapproval.

The flow of weapons to the Home Guard had increased and the winter months offered opportunities for training on new weapons in drill halls, scout halls, church halls and all the other buildings where the new force had found shelter. The continuing shortage of .300 rifle ammunition caused problems in training and much of the range work had to be done with the few .303 rifles units had been allowed to retain. The 2nd Angus Battalion had received the welcome delivery of 400 American .300 rifles in mid-July but had only been supplied with 8,000 rounds of ammunition. Fifty rounds per rifle may at first glance seem a substantial allocation but these 400 weapons were being shared between 1,300 men. The amount of firing practice that any man could get was therefore quite limited and the shortage of ammunition would have been more than embarrassing had the unit been called to go to action stations.

Often the in-house training effort could be aided by informal assistance from army or other units stationed in the area. The 2nd Renfrewshire Battalion in September had appointed Major Morrison to take charge of training on automatic weapons. After he had himself been trained on the Browning automatic rifle he set up a battalion training course with four men from each company detailed to attend. In November a Royal Marine unit stationed in Paisley offered an officer and two other ranks to assist the battalion on Vickers and Lewis machine-gun training.

At the end of the year the Home Guard could look back with pride and some wonder on its accomplishments and at what had been achieved from a standing start. Almost by accident something had been created which with its dispersed structure and its "footprint" in every corner of the land, was well designed to deal with the threat of *blitzkrieg*. France had found that once the front line had been breached the German tactics of infiltration and attack in depth had not been easy to resist. John Brophy had identified this problem: "Part of the weakness of France, confronted with invasion, came from the fact that so many men, of all ages, had been 'called up' and embodied in the large formations of the regular army. The towns and villages, behind a line which did not hold, were denuded of the means of local resistance. The Home Guard network ensures that this mistake will not be made here."[35]

The growing strength of the Home Guard dispersed throughout the land and potentially fighting on ground they knew well represented a significant complementary force to that of the field army. The familiarity with the territory on which they would be expected to fight was a huge asset for the Home Guard. Major Sellar, of the 2nd Perthshire made this point: "A few months ago a battalion [of the Home Guard] was given a brief course of instruction in mountain-warfare, and then a brigade of regulars with artillery and all accessories was launched against them. After 12 hours of the most realistic fighting, the attackers had made little or no progress against the local men who knew every rock and ridge for miles around."[36]

Indeed one of the less obvious roles of the Home Guard was the provision of sections of guides, men like farmers, foresters and gamekeepers, intimately familiar with the local countryside, who would be attached to regular units operating in the area and assist them in moving swiftly and effectively across country. In this way the local knowledge and skills of the Home Guard could be used to the advantage of the field Army. The history of the 3rd Renfrewshire Battalion relates how their guide section under the battalion guide officer assisted a unit of Grenadier Guards in a night exercise on Fereneze Braes and "performed their duties in a manner which resulted in very high praise from the Commanding Officer of the Guards."[37]

There was no merit in turning the Home Guard into another army, however much some of the more ambitious Home Guardsmen may have yearned for this. It lacked the supporting arms like artillery and engineers and the command structures to be used as a normal field force in brigade, division or corps strength. However its true strength and unique significance were not as a second-grade field army but as a locally rooted force, defending their own homes, fighting with home advantage, and as such the Home Guard had, almost inadvertently and in a remarkably short time, become a force to be reckoned with. The commander of the Tannadice Platoon of the 2nd Angus Battalion wrote in his operational orders of the need to maintain strong forces to strike at the flanks and rear of the enemy; reserves would not, as in the First World War, be thrown in to reinforce the front line because: "In this War there is no "front" in the accepted sense. This has been brought about by the enemy tactics known as "infiltration"; by his use of Air-Borne Troops and by his methods of employing fast motorised columns."[38] Had this requirement for defence in depth been required to be met by the regular army alone there would have been little prospect of freeing enough regular troops for the campaigns in North Africa and the Middle East.

Anthony Eden, winding up a debate on the Home Guard in the House of Commons on 19 November, said: "No one will claim for the Home Guard that it is a miracle of organisation ... but many would claim that it is a miracle of improvisation, and in that way it does express the particular genius of our people. If it has succeeded, as I think it has, it has been due to the spirit of the land and of the men in the Home Guard."[39]

The Secretary of State acknowledged the continuing threat of invasion and went on to look forward to the winter and to the year that was to come – the campaign of 1941. In a sentence which marked quite how far the Home Guard had come since his May broadcast creating the Local Defence Volunteers, he said: "It is clear that the Home Guard could never act wholly as a substitute for the field Army, and it would not be right to try to encourage it to do so; but it can be an auxiliary so valuable as to release important elements of the field Army to go and fight elsewhere, and that is the role which we hope to see the Home Guards play in 1941."[40]

Chapter 6

ARMING THE GUARD

In the four years of its existence the Home Guard went from the improvisatory stage of enthusiastic amateurs drilling with broom handles to which kitchen knives had been lashed – and, if they were lucky, donated shotguns – to being a formidable and well trained force armed with a huge quantity and great variety of weapons. This chapter will look briefly at the range of infantry weapons issued to the Home Guard; the force's involvement with coastal and anti-aircraft artillery will be looked at in later chapters.

A typical Home Guard general service battalion such as the 2nd Aberdeenshire, operating in the Huntly, Alford and Inverurie area of central Aberdeenshire had, at stand-down in 1944, 1,862 men armed with the following very considerable and varied arsenal:

Rifles .300	1,008
Rifles .303	28
Rifles 0.55 anti-tank	12
Sten guns	602
Browning automatic rifles	50
Lewis light machine-guns	18

Vickers medium machine-guns	2
Browning medium machine-guns	6
Spigot mortars	12
Smith guns	14
2-pounder anti-tank guns	4

Nor was this the totality of the weapons issued to the Home Guard. Revolvers were issued, as was the grenade-launching EY rifle and the Northover projector. The latter device, exclusive to the Home Guard, fired standard hand grenades up to 200 yards. There was also, of course, a range of grenades for hand throwing.

Weapons were more than just a technical issue – they were major elements in the building of morale. While there would have been an undoubted administrative gain from a simplification of the range of weapons there was a very clearly perceived advantage in morale terms in keeping the Home Guard interested and alert by providing a steady stream of new weapons which had to be mastered and then incorporated into the force's ever-changing operational role. The regular army expected to be trained and re-trained to cope with changing demands; the 52nd (Lowland) Division spent two years training as a mountain warfare unit, then re-trained as an air-landing division, but never fought in either role, eventually itself fighting in the distinctly unmountainous Netherlands. All this training and re-training was undoubtedly irksome but at least there was a reasonable expectation that ultimately it would not be wasted and that in some form or other the men of the division would see service. The morale problem of the Home Guard was a more acute one. The Home Guard had to remain alert and efficient, to undertake strenuous and at times tedious training against a background of a steadily decreasing threat of invasion. The wave of patriotic enthusiasm and devotion to duty which had attended its birth could only be seen as being likely to decrease as the perceived threat receded. The less likely a guardsman was to see the German army arriving on his doorstep the more he would be inclined to wonder whether it was worth turning out for drills and parades on a cold winter night. Although the terms of service of the Home Guard would, as will be discussed in the next chapter, change in 1942 it remained throughout an unpaid service. This, and the original voluntary constitution and "equality of service" ethos required careful management of the force and a degree of sensitivity to maintaining its commitment and enthusiasm, a sensitivity that would probably have not been displayed towards the regular army. For example, in November 1942 Scottish Command ordered that one of the Home Guard's artillery weapons, the Smith gun, was in future to be employed only in large towns and important focal points where there was a garrison large enough to make a protracted defence. The instruction went on: "In order that disappointment may not result from this withdrawal efforts should be made to replace by some other weapon, such as the Spigot Mortar."[1]

From July 1940 onwards the stocks of .303 rifles that had been issued were generally withdrawn, although, as the weapons inventory of the 2nd Aberdeenshire shows, a small quantity of these rifles were retained for target practice due to the much greater availability of .303 ammunition. Other than this specific use there was a general movement towards the issue of .300 calibre US-manufactured rifles; over the same months stocks of Lewis guns, Browning automatic rifles and some Vickers medium machine-guns found their way to Home Guard units. In Glasgow a separate machine-gun unit of company strength was created in 1940 and held as a reserve unit at the disposal of the local garrison commander.

The 12-bore shotgun continued to feature in the Home

Guard armoury, and when used with a solid ball cartridge was seen as a useful short-range weapon even in the hands of less than highly trained men. It was considered to be particularly useful in a confused town-fighting situation where the long range of the rifle was not only unnecessary but could be dangerous to friendly forces and civilians.

One of the instructors at the Osterley Park Home Guard Training Centre, and who later transferred to the official War Office school at Denbies, Surrey, was Albert "Yank" Levy, a specialist in unorthodox warfare and a former member of the International Brigade who had served in Spain under Tom Wintringham. In January 1941 160 men of the 6th Perthshire Battalion attended a lecture given by Levy on "Irregular Warfare and Guerrilla Tactics" and they particularly enjoyed his suggestions on unorthodox ways of killing Germans, for example by use of a knitting needle.[2] "Yank" Levy was by birth Canadian, but had grown up in Ohio and had fought in a number of trouble spots around the world. He wrote a manual on guerrilla warfare aimed at the Home Guard which was published in 1941 by Penguin Books and enjoyed a large sale, like so many other Home Guard "self-help" books.

Reference has already been made to the Molotov cocktail, the improvised anti-tank weapon developed in Finland; however, more conventional grenades were also supplied to the Home Guard. These included the normal type No. 36 grenade, often known as the Mills bomb, as well as No. 76 SIP smoke grenades and No. 68 anti-tank grenades. Unfortunately the grenade proved to be the most dangerous of the Home Guard's weapons. The death of Major F.D. Mirrielees and Volunteer D. McNiven in Perthshire was referred to in the previous chapter and other grenade training accidents would occur quite often. In August 1942 a regular army officer, Major Ralph Joynson of the Black Watch, who was serving with the Home Guard as a training officer, died in the military hospital at Inveraray from wounds sustained in a grenade accident in Argyllshire. A company commander of the 1st Argyllshire Battalion, Major L. Methuen Campbell, was injured in the same incident. Another officer of the 1st Argyllshire, Lieutenant Downie, was injured in a grenade accident in 1943.

The 2nd Renfrewshire Battalion acknowledged the problem and the real danger it posed. They decided to adopt a policy of training only 25 per cent of their men on grenades but to train them well, rather than training all of the men to what would inevitably have been a less high standard. It was this unit's view that not all Home Guardsmen were temperamentally suited to the use of the grenade, that the infantryman's basic weapon was the rifle, and that the very real and widespread fear of accidents caused a loss of confidence and itself increased the risk of mishaps. The battalion's engineer section built a bombing pit at Hawkshead, near Paisley. Men of the 2nd and 4th Renfrewshire Battalions were trained in live grenade throwing there and used the site without suffering any accidents. This training facility was also made available to regular forces stationed in the area. Sadly the 3rd Renfrewshire were less fortunate in their bombing practice and had two officers injured in grenade accidents.

In the spring of 1941 the first supplies of the undoubtedly peculiar looking Northover projector found their way to Home Guard battalions, Glasgow Zone being issued with 36 of them in April with additional supplies coming through over the next year. This weapon resembled a piece of drainpipe on four legs and used a small explosive charge to propel a grenade considerably further than a man could throw it. The writer John Brophy commented: "This piece of sub-artillery is at least ten times as good as it looks at first sight, and the simplicity of its operation should not blind anyone to its effectiveness." He went on to note: "The accuracy of the Northover, in view of the fact that it has an unrifled barrel, is astonishing, and it can be mastered with a remarkable ease and speed."[3]

In Glasgow the first supplies were allocated to perimeter defence battalions but eventually all battalions in the city received supplies and set up what were called "sub-artillery" sections or platoons to operate the new weapon. The Northover projector may have looked odd but it did offer the Home Guard a useful weapon. The men of the 3rd Stirlingshire Battalion were distinctly enthusiastic about the new device: "Officers and NCOs were taken to No. 2 War Office School at Kinnaird House to see the Northover Projector demonstrated. Some Officers were fortunate enough to see the demonstration carried out by the inventor himself, Major Northover, who succeeded in putting across a splendid demonstration in a most humorous fashion. So much so that complete confidence was instilled in the men regarding the possibilities of the weapon."[4]

The 1943 War Office Home Guard Instruction Manual No. 51 described the tactical use of the Northover Projector as: "...an ambush anti-tank weapon to be sited as part of the defences to fire enfilade on likely tank approaches. As it is easily transported a commander can allot a projector and its detachment to a standing patrol needing additional anti-tank defence."[5]

In addition to the anti-tank role it could be used to fire incendiary grenades to provide smoke cover for an attack, although the smoke produced from the SIP grenade had toxic properties and could not be used if friendly forces were to remain in the smoke screen area for an extended period.

In the winter of 1941–42 the next Home Guard weapon came into service. This was the 29mm spigot mortar, sometimes known after its inventor, Lieutenant-Colonel Blacker, as the Blacker Bombard. (The 29mm measurement referred to the bore of the spigot or rod which, after the firing of an explosive charge, propelled the bomb from the mortar.) Originally designed for issue to the regular army, it was overtaken by the development of the much lighter PIAT (Projector Infantry Anti-Tank) and the spigot mortar was issued instead to the Home Guard. The Blacker Bombard fired a heavy 20lb finned anti-tank bomb to a range of over 100 yards and with that weight of explosive and range it was an effective anti-tank weapon. An alternative 14lb anti-personnel round was available and had a range of 500 yards. An inert concrete shell could be used for training purposes and this proved useful as some Home Guard units found difficulty in obtaining or creating a firing range where these new sub-artillery weapons could be exercised. The Bombard was a heavy and cumbersome weapon, weighing 350lb, which required a large crew to move; this was not of course a major issue in the manpower-rich Home Guard. It could also be semi-permanently mounted on a concrete block as part of a defence system; the 1st Sutherland Battalion which was charged with the defence of the Bonar Bridge locality in 1942 used most of its allocation of spigot mortars and Smith guns in this task.

The official doctrine on the spigot mortar was that it was the most destructive of the Home Guard anti-tank weapons but had problems with a slow rate of fire and even if camouflaged was fairly easily spotted and neutralised once it came into action. However "a direct hit will almost certainly severely damage, if not destroy, any heavy tank now known."[6]

The Smith gun, which was mentioned above, was a privately developed emergency artillery piece invented by William Smith, the chief engineer of the Trianco Engineering Company in Sheffield, and was designed by him to meet the urgent post-Dunkirk shortage of artillery. It took a 3-inch mortar shell and could fire the anti-tank version 200 yards and the anti-personnel version

650 yards. It was about as simple an artillery piece as could be envisaged – a smooth bore barrel and a carriage consisting of two metal wheels. When it went into action the gun was turned over onto the right-hand wheel, with the other wheel providing a degree of overhead protection. Operated by a gun commander and three men, the Smith gun or OSB (Ordnance Smooth Bore) gun could be manhandled into position and turned on its side by the detachment. It was never issued to the regular army but Home Guard units received substantial quantities, with Glasgow Zone taking delivery of 25 guns in February and March 1943 and a further batch of 12 guns in October that year. As with the spigot mortar, many battalions had difficulty finding a live-firing range, the 2nd Renfrewshire, for example, were allocated eight Smith guns in February 1943 but could not conduct live-firing drills until the summer of 1944.

The official manual claimed that the Smith gun could penetrate 80mm of armour plate at 50 yards. A crew would need good nerves and good training to hold their fire until an enemy tank approached to within 50 yards. The Smith gun was, however, "very mobile and could be towed behind a 10 h.p. car or a motor cycle."[7] However, the weapon had its limitations, as a conference on Home Guard training in July 1942 heard. The gun could not be dug in to fire, and cross-country transport was problematic as on rough ground the tyres tended to come off the wheel. Although the design was intended to provide some protection, neither the shield nor the wheels were bullet-proof.[8]

Despite their shortcomings these sub-artillery weapons represented a considerable resource for the Home Guard, and their simplicity of construction and comparative cheapness meant that they could be made available in very substantial quantities. The 3rd Scottish Borders Battalion, based in Berwickshire and with a significant coastal area to defend, had by 1944 no fewer than 97 spigot mortars and 20 Smith guns in its inventory. It was considered easier to train the part-time volunteers of the Home Guard on a weapon such as the Northover projector or the Smith gun than on more sophisticated if more effective weapons, which in any event were in demand for the field army.

In September 1943 a meeting with General Lord Bridgeman, Director General Home Guard and senior staff officers from Home Forces and the War Office agreed that the Home Guard should retain spigot mortars and Smith guns, that 2-pounder anti-tank guns and 0.55-inch anti-tank rifles should be issued when available, and that the Northover projector should be replaced by the PIAT when supplies became available.[9]

Later some Scottish units received supplies of the 2-pounder anti-tank guns which had been issued to the regular army but were gradually replaced by heavier calibre anti-tank weapons in the Royal Artillery anti-tank regiments. The 2-pounders were issued to infantry battalions to arm anti-tank platoons and some were cascaded down to the Home Guard. Although there were serious limitations to its effectiveness against heavy armour, the 2-pounder was accurate, quick-firing and had a longer useful range than the Home Guard's existing sub-artillery weapons; it was most effective up to a range of 200 yards but could be used up to 500 yards. The rather ambitiously titled anti-tank rifle or Boys rifle (named after Captain H.C. Boys of the British Small Arms Committee), firing a 0.55-inch bullet, was also issued to Home Guard units. The instruction manual was reasonably frank about its limitations: "The A-tk rifle is *not* effective against heavy armour. . .It can, however, break the tracks of heavy tanks. The bullet can pierce light armour and put soft vehicles out of action."[10] The Northover projector remained in the inventory of Scottish Home Guard units until stand-down and few PIATs seem to have found their way to Scottish battalions.

Machine-guns of various patterns were also found in the Home Guard arsenal; the Vickers medium machine-gun was a standard weapon for both the regular army and the Home Guard. However, the most common machine-gun in Home Guard service was the Lewis gun, a weapon dating back to the First World War and firing the US .300 calibre round. Many Lewis guns were imported from the United States in 1940 to give the Home Guard some much-needed firepower. The 1st Sutherland Battalion deployed 10 Vickers machine-guns and 30 Lewis guns.

Part of the US-manufactured inventory of the Home Guard was the effective Browning automatic rifle which fired .300 calibre bullets from a 20-round magazine and was widely issued to provide heavy fire support at platoon and squad level. This was a popular weapon of obvious quality. The American Thompson submachine-guns, the tommy-guns beloved of Chicago gangsters, firing a heavy .45 round, were also issued in small numbers in 1940 but did not become a significant part of the Home Guard arsenal.

The commonest Home Guard weapon after the rifle was the Sten gun. This 9mm submachine-gun had the great advantage of being an extremely basic and unsophisticated design which could be turned out very cheaply; indeed it was reputed to have cost a mere £1.15.0 (£1.75) to manufacture. Most Home Guard battalions ended the war with several hundred Sten guns in their inventory. The weapon, although only accurate at short ranges of up to 50 yards, was found to be ideally suited for town fighting and short-range skirmishing in woods and because of its compactness it was also issued as a personal weapon to drivers of Home Guard transport units. It was simple to use and required little in the way of theoretical training. The light Sten gun (the name was an acronym based on the initial letters of its designers' names – Shepherd and Turpin – and EN from Enfield, the site of the

Royal Small Arms Factory where it was developed) weighed only 3.18 kg. This was less than the 3.95kg of the Springfield rifle and significantly less than the 8kg of the Browning automatic rifle or the extraordinarily heavy Boys rifle which weighed in at 16kg and measured a cumbersome 1.575 metres in length. The Sten used 9mm ammunition, a common continental size for pistols, and could thus use captured German ammunition; it fired these cartridges at a rate of up to 500 rounds a minute using a 32-round box magazine.

By the time thousands of Sten guns and all the weird and wonderful forms of sub-artillery had been issued, the Home Guard was able, with some justice, to see itself as a well-armed, well-trained and effective force. It was told repeatedly how important it was to the nation's defences; it was, in Churchill's words, as much part of the British army as the Grenadier Guards. Its presence allowed the field army to send large formations to the Far East and the Middle East, secure in the knowledge that the Home Guard was taking part of the strain of guarding the homeland against invasion or attack.

In this context the blow to the Home Guard's morale caused by the inept decision to issue a weapon which seemed to take them back to the pitchfork and broom-handle days of May 1940 cannot be underestimated. In 1941 Home Guard numbers had exceeded arms available and a directive had come down from Churchill that every member of the Home Guard should have a personal weapon. In itself this was an entirely sensible instruction and one that, correctly implemented, should have been conducive to improved morale; as the historian of the 3rd Stirlingshire wrote: ". . .probably the greatest day of all was when every man in the battalion was issued with a rifle or Sten Gun of his own, with the knowledge that there were ample supplies of ammunition."[11]

However in the summer of 1941 there was still a shortage of

such weapons and somewhere in the less imaginative recesses of the War Office the idea came of issuing a weapon whose last triumph had been in the armies of Marlborough in the 18th century and which, as one Scots officer observed, had been found sadly wanting at Culloden. The pike was reinvented, in the guise of a length of steel tubing with a bayonet spot-welded to the end. The "bayonet-standard", as the War Office preferred to call it, was more or less universally seen by the Home Guard as an insult; unit historians writing more than two years later were still clearly incensed by the blunder of issuing this implement. The Home Guard needed to be handled with some discretion – they were unpaid volunteers, after all – and the reintroduction of the pike, along with the unconvincing official message that it was ideally suited to street fighting, made many members feel that they were not being taken entirely seriously. Most commanding officers refused to issue the pikes to their men; in the 3rd Stirlingshire the men themselves refused to accept them and the history of the 7th City of Edinburgh Battalion notes with evident irony that 100 pikes were delivered to the battalion but: ". . . in order to prevent homicidal tendencies from developing with the aggressive spirit in the Battalion, these were never issued."[12]

Had the pikes been issued when the Local Defence Volunteers were formed, in the crisis months of May and June 1940, then they might have been welcomed as a slightly more sophisticated version of the kitchen knife lashed to a broom handle, but after years of training with an ever widening variety of modern weapons, the idea of parading with pikes was anathema to the officers and men of the Home Guard. The problem should have been recognised within the Home Guard directorate, who, to be fair, were usually more aware of the sensitivities of their millions of volunteers.

The pike issue provoked much public comment, press criticism and debate in parliament. The Earl of Mansfield, a major in the 2nd Perthshire Battalion, observed: "... frankly the Home Guard honestly regard these pikes as little less than an insult. If they have been supplied in the same proportion throughout the country as they have been to that battalion in which I serve, I estimate that already not less than 1,000 tons of valuable iron and steel have been wasted in this way, and that, at the present time, I think, is little short of deplorable."[13]

It should be remembered that at this period park railings and ornamental ironwork were being chopped up for salvage drives and housewives were being encouraged to donate spare saucepans to build aircraft. Lord Croft, Joint Parliamentary Under Secretary of State, enunciated the official line on the pike in the less than persuasive words: "I venture to think that if you have a bayonet it is a useful weapon, and if that bayonet is lengthened by a staff or stave it is a still more useful weapon."[14]

Although the weight of opinion was against the pike, at least one writer to the *Scotsman*, Douglas G.W. Wilson, a retired army officer, put a positive spin on the matter when he wrote: "... if I were given my choice between being armed with a pike and being armed with the promise of an anti-tank rifle, I, personally, would prefer a pike every time (unless I could beg, borrow, or cadge a claymore)... A dead German is a dead German, even if he has been killed awkwardly with some antiquated weapon."[15] Which was true, but hardly explained how the German in his tank was to be killed either by pike or by claymore. The author of the 1st West Lothian Battalion history probably expressed a common attitude when, after explaining that his battalion's allocation of pikes had not been issued, wrote: "it was early realised their value was problematical and their effect on the morale of the men likely to be detrimental."[16]

Chapter 7

THE LONG HAUL 1941–43

The story of the Home Guard from its inception as the Local Defence Volunteers to the end of its first year of life in December 1940 has been covered in earlier chapters. By the start of 1941 the most immediate threat of invasion had receded. The failure of the Luftwaffe to win air supremacy by defeating the Royal Air Force in the Battle of Britain in the late summer and early autumn of 1940 and the continuing naval superiority of the Royal Navy – a dominance underlined by the destruction of the French fleet at Oran in June 1940 – made a German sea-borne invasion, which was totally dependant on the navy being neutralised and air superiority being attained over the Channel, impracticable. Accordingly the German plans – Operation Sealion – were abandoned in September 1940. An invasion was thought to be impossible in winter weather and when in June 1941 Germany launched Operation Barbarossa – the invasion of the Soviet Union – the likelihood of Germany being able to mount an invasion of Britain in 1941 seemed slight. Of course had Barbarossa succeeded and the Soviet Union been eliminated then German attention could again have turned to the west and the defeat of Britain, the only remaining power resisting Nazi hegemony. Although full-scale invasion was a decreasing risk there was at all times the danger of raids or small-scale landings, sabotage and attacks aimed at shaking civilian morale and destroying important installations and facilities.

The enthusiasm fuelled by invasion fears, which had brought 1.7 million men into the Home Guard by January 1941, might naturally be expected to dwindle as the prospects of German invasion receded. The remarkable fact is that the Home Guard continued to attract the loyalty, support and energy of millions of men over the next four years. This chapter will look at the period from 1941 to 1943 and the changing role of the Home Guard, which was not, on a number of counts, the same organisation at the end of this period as at the beginning. Its focus will be on the general service battalions, the infantry of the force, as the role of the Home Guard in artillery and other specialised areas will be dealt with separately.

In the previous chapter the part that new equipment played in maintaining Home Guard morale was emphasised. It is equally true that the changing operational role of the Home Guard was also of real importance in retaining the interest and commitment of the men and it is remarkable how well that commitment was maintained. In January and February of 1943 the 1st Edinburgh Battalion maintained a covert check on attendance at parades and found that an average attendance of over 73 per cent was recorded; as this figure made no allowance for sickness or authorised absence it is a very high attendance rate and surely indicative of the high state of morale in the unit.

It was always a major concern of senior officers and politicians to keep the Home Guard motivated, especially as the war progressed. In July 1943 Lord Croft, the Under Secretary of State for War, did his bit in this cause when he told a Home Guard parade that soon large parts of the country would be relying on them for defence. He meant that as regular troops were sent overseas and the long-awaited Second Front opened up, the country would be stripped of regular formations. Despite the improving military situation Lord Croft said: "Hitler, the desperate mad dog, may launch an attack which, if it came, would be by air borne units and the Home Guard would then be of greater value than ever before."[1]

From the LDV's initial role of watching and reporting on enemy parachutists and static defence of the local area, the Home Guard, as training and equipment developed, diversified into a whole range of functions. The historian of the 3rd Stirlingshire Battalion gives an insight into some of these changes: "Training was at all times governed by directives from higher authority, which detailed the role of the Home Guard. The role changed periodically from Guerrilla Warfare, Static Defence, Anti-Para-troop Defence, Ambushes, Offensive Defence, etc. In addition some units were specially trained in observing and reporting, and in how to deal with Fifth Columnists and Saboteurs. The latter required little instruction, and the personnel were quite clear in their own minds as to how Fifth Columnists and Saboteurs should be dealt with."[2]

Although this changing nature of training was undoubtedly a reality it is important to remember that a Home Guard unit was at any given time only tasked with one function. The Director General of the Home Guard, Major-General Lord Bridgeman, expressed the similarity and the difference between the regulars and the Home Guard when he said: "The difference between the Home Guard and the Regular armed forces of the Crown was that the Home Guard was practising for only one battle. This was the battle each company, section or platoon would be fighting on its own ground."[3]

One of the problems that inevitably faced the Home Guard was the varying degree of fitness and physical ability among its members. An army infantry battalion was a much more homogeneous body, consisting of young men in good physical condition who could be expected to march substantial distances with heavy equipment and fight at the end of their march. Even the senior officers of an infantry battalion were, by the middle of the war, likely to be in their 30s. The Home Guard, of course, included young and fit men in reserved occupations who were, in theory, up to every physical challenge. It also included physically immature teenagers and men in middle age and old age who might have difficulty carrying a combat infantryman's equipment, let alone completing an assault course. A field army battalion was expected to take part in modern mobile warfare, to fill a variety of roles, to co-operate with other arms and services; the Home Guard battalion did not have the infrastructure or the personnel for such a form of warfare, and despite the considerable enthusiasm of the guardsmen there was little sense in attempting to fit its particular round peg into the square hole of normal military practice. The 3rd Dunbartonshire Battalion history reflected that: "The average Home Guard was past his athletic prime and could function best in the static role. Training was based accordingly and with the young men on reconnaissance work and other mobile jobs."[4]

There were various suggestions that the Home Guard could be made into a mobile force. One of these originated with a minute from Winston Churchill in November 1941 suggesting that four mobile battalions be created in each corps area. The War Office response noted that Home Guards were trained to do one task – to defend their home area as part of the local defence scheme – and that to implement the prime minister's scheme would involve moving men away from their homes, demand full-time service and have implications for organising and equipping these units on a war footing comparable to the regular army. One mobile Home Guard formation was created within Scottish Command: 14th City of Glasgow

(STC) Battalion, which drew its manpower from the University Senior Traing Corps.[5]

As a more active role began to be considered for the Home Guard it was recognised that not all members would be able to undertake all duties and a formalisation of what had always been a pragmatic response to the varying capacities of the volunteers came into effect. A company commander in the 6th Ayrshire Battalion, Captain B. Knox, developed the concept of mobile platoons, which would utilise the younger and fitter members of the Home Guard in a more active role. He sent this idea up through the chain of command and it was soon adopted widely in the Home Guard. The older or less fit members would undertake static duties, such as guarding vulnerable points or manning observation posts.

Thus the company-sized Tannadice Platoon of the 2nd Angus Battalion was, as described in Chapter 5, divided into three static sub-platoons and two mobile battle sub-platoons. There was a role for the oldest and least fit members in providing HQ office and orderly functions. Detailed plans were made for implementation in the case of an invasion or raid; the platoon headquarters would be activated and arrangements made for a designated individual to control the petrol pump at Finavon Hotel. From here petrol would be issued to Home Guard vehicles and preparations made for the contamination of the petrol supply with sugar or water to deny it to the enemy. Because the Home Guard when called to action stations would become a full-time service, arrangements had to be made for pay and rations. The manse of Oathlaw became not only the platoon first aid post, with the minister, Private A.J. Oliphant, as ambulance corporal, but also housed the ration centre and pay office and was the base for the Quartermaster Corporal and the Pay Corporal under the direction of the Admin Officer, 2nd Lieutenant J. Lee.[6]

The Home Guard's strength lay in its unrivalled knowledge of the local area and even the best-trained regular troops could come off second best when matched against a Home Guard unit on its home turf. The account of the 6th Ayrshire Battalion's progress over these years underlines this fact:

> In the earlier years there were several exercises against Commando units mainly in the Dalmellington area which produced a rough-house at the time, followed by friendship and much assistance; after having given us a practical demonstration on the value of unarmed combat they sent out instructors to teach us how it was done. During 1943 C Company had quite a lot of exercises with air force parachutists trying to make their way in secretly through our area to test the aerodrome defences and became very expert at catching them, especially after they had developed a special technique of cycle patrols. This principle was later adopted throughout the Battalion both for recce. work and for rapid movement of fighting patrols. Finally in early 1944 all companies had most interesting and valuable exercises with some paratroop units and also a special highly experienced sabotage unit who were in the area; these exercises took place both by day and by night and provided exactly the type of operation in which at that stage of the war the Home Guard seemed most likely to be involved. The results showed how easily a sentry could be put off his guard by experts and then caught napping; on the other hand they also showed the value of local knowledge and that, given the right opportunity, the Home Guard were often able to lay even experts by the heels.[7]

As the extract above suggests the Home Guard had from its earliest times trained hard and a variety of centres were soon developed to provide expert instruction in a variety of skills. The early unofficial training centre at Osterley Park, replaced by an official school at Denbies, was swiftly replicated in Scotland with No. 2 War Office School being established at Kinnaird House, Larbert, near Stirling in 1941 and a Home Guard Tactical School which was set up at Moncrieff House, Bridge of Earn. Among the subjects dealt with at Kinnaird House was the ever-popular topic of ambushes. The local Home Guard battalion, the 3rd Stirlingshire, which conveniently included many countrymen with great experience in such occasionally useful rural skills as poaching, was asked to provide the demonstration party every week for many months, a task which pleased the men of 3rd Stirlingshire and aided the development of their own skills.

Kinnaird House and Moncrieff House were centrally organised schools, and battalions from all over Scotland sent officers and non-commissioned officers there, as well as to the Scottish Command Weapons Training School at Edinburgh, to be trained to become trainers. However in many areas local training provision, often of a very elaborate nature, was developed.

The initial call to the LDV had excluded cities and large urban centres but, as we have seen, this policy was swiftly abandoned and a need for specialised training in urban warfare in cities like Glasgow and Edinburgh was quickly recognised. Large-scale exercises were conducted; for example, in July and August 1941 exercises involving over a million Home Guards across Britain were arranged, with 10,000 Home Guardsmen defending Glasgow. Regular troops took on the role of the "invader" and umpires agreed that the Home Guard had distinguished itself and beaten off most attacks. An impression of the exercise is revealed in a *Glasgow Herald* report: "So realistic of battle was the scene outside and inside an important railway centre in Glasgow yesterday forenoon that a large crowd of citizens who had been drawn to the neighbourhood by the sound of crashing bombs, the bursting of grenades, the rat-tat-tat of tommy guns, and the actual smashing of windows thought that something serious was afoot."[8]

Wartime censorship did not permit the naming of the railway station, which was in fact the now-vanished Buchanan Street Station. It was defended by a company of the 7th City of Glasgow (LMS Railway) Battalion and attacked by a troop from No. 6 Commando. The Buchanan Street Station complex was both the HQ of Glasgow Home Guard Group II, the 7th Battalion, and was a recognised Vital Point in respect of the LMS Control and Telegraph Centres located there.

Perhaps as a result of exercises like this, Glasgow Zone decided to create its own Town Fighting School in a row of ruined houses at Dixon's Blazes off Cathcart Road in the south side of the city. "Glasgow's well-known thoroughfare, the Gallowgate, has been reproduced in paint, complete with the names of famous 'howffs', on the skeleton walls of what was once a miners' row and is now the main thoroughfare of the Glasgow Area Town Fighting School. Here, every week-end, companies of Home Guardsmen are taught the art of house-to-house fighting. The methods adopted, the majority of which are peculiar to the school, instil into the students all the tricks, cunning and craft of the guerrilla fighters."[9]

The chief instructor at the Glasgow school was Captain Thouron, succeeded by Captain A.B. McLetchie, assisted by officers from the 2nd, 3rd and 11th City of Glasgow Battalions. The syllabus included demonstrations and live firing of grenades, bombs and Northover projectors. The Dixon's Blazes school, which was recognised as one of the best in the UK, was also used for the training of regular army troops and US forces. A similar school was created in Couper Street, Leith, for the East of Scotland and an Aberdeen Town Fighting School, established in derelict properties in the Footdee area of the city, also flourished in 1942. A report on the Couper Street School, referred to in the censored press of the period as "a Street Fighting School in the South-East of Scotland," enthused about:

> . . . the keenness of the instructors and squads in getting the utmost advantage from this special course. At the end of the course the officers and men return to their battalions to pass on the new training. . .
>
> It is training that concerns all ranks and all arms, and the point is made that street fighting is especially a soldier's battle. As one of the officers put it, "the troops can be shown how to avoid many of the more elementary forms of suicide. . ."
>
> The demonstrations took place in an area in which there is a considerable amount of dilapidated property, and it formed an ideal background for the mock battles of the day, in which rifles and anti-tank guns were fired and grenades thrown.[10]

Lord Bridgeman, Director General of the Home Guard, visiting Scotland in May 1942, spoke with enthusiasm of the development of street-fighting training which, he said, "was catching on like wild-fire" and he commended the varying approaches to urban warfare training which he had seen at Glasgow and Edinburgh.[11]

In addition to training centres and the regular training carried out by battalion staff, there were mobile training teams sent out from the War Office. One of these teams was in the Kintyre and Islay area of the 3rd Argyllshire Battalion for ten days in February/March 1943 and trained local members in the use of Spigot mortars and grenades, as well as in matters such as battle drill and fieldcraft. These travelling teams of officers, warrant officers and NCOs went to all parts of the country, with sessions conducted in places such as Carradale and Southend in Kintyre and Bridgend on Islay.

However, even the best-laid plans of the Home Guard could be thrown out by unforeseen problems. In November 1941 D Company of the 5th City of Glasgow Battalion had arranged to defend a Govan foundry against Special Service troops of the regular army – what would later be called commandos. Major J.M. Wilson, the Home Guard Company Commander, later told the *Glasgow Herald* that the exercise had been very satisfactory but his defending force had not been at full strength as many members had been unable to reach the exercise area due to a bus strike.

The role of the Home Guard developed and a policy of "offensive defence" came into play. As battle platoons began to form, staffed by the younger and fitter members, so training facilities developed all over the country to equip the Home Guardsmen for this new role. The Angus area had a Guerrilla School at Cortachy with an officer from the 2nd Angus Battalion, 2nd Lieutenant Stewart MacFarlane, as chief instructor. This officer was a good example of the Home Guard's ability to find its own instructional staff. MacFarlane, in civil life a blacksmith from Finavon, was promoted corporal in July 1941, sergeant in February 1942 and commissioned in May 1943. The 1st Moray Battalion established the North Highland District Active Service School, at Humbreck Farm, south of Elgin. This ran for several weeks in the summer of 1941 and provided around thirty men at a time with a weekend of active-service type experience. In 1942 the 1st Moray created within each of their companies a battle platoon structured as follows:

Above. Perth Home Guard. Note that chevrons appear on one arm only – in 1941 when non-commissioned ranks were introduced they became worn on both arms (Thomas Sinclair)

Left. Officers of the Singer's Company, 2nd Battalion Dunbartonshire Home Guard in front of the famous Singer's Clock – emblem of the Singer Sewing Machine Company (Clydebank Libraries)

Home Guard Patrol at Loch Stack, Sutherland (Imperial War Museum)

Home Guard manning Lewis Gun at Loch Stack, Sutherland (Imperial War Museum)

Smith Gun or OSB 3" Gun (Imperial War Museum)

The Northover Projector (Imperial War Museum)

Right. Blacker Bombard or Spigot Mortar (Imperial War Museum)

Below. Home Guard Exercise at Glasgow Railway Station (*Daily Record*)

Above. Glasgow Town Fighting School (Imperial War Museum)

Left. Glasgow Town Fighting School (Imperial War Museum)

Right. Proficiency Certificate for James Walker (James Walker)

Below. Callander Home Guard Pipers (K. Dunn)

Bottom. Home Guard guarding Hess's aircraft (*Daily Record*)

Above. Z Rocket Firing Demonstration for Scottish newspapermen (Imperial War Museum)

Left. Canal Patrol, Union Canal, Edinburgh (Imperial War Museum)

Top. Beaverette Armoured Car in Highlands – named after Lord Beaverbrook, Minister of Aircraft Production (Imperial War Museum)

Above. Men of the Yarrow's Company 4th (City of Glasgow) Battalion demonstrating the range of infantry weapons used by the Home Guard (BAE Systems)

Left. Service Certificate (Thomas Sinclair)

In the years when our Country was in mortal danger

THOMAS SINCLAIR,

who served from 24 Mar 41 to 31 Dec 44, gave generously of his time and powers to make himself ready for her defence by force of arms and with his life if need be.

George R.I.

THE HOME GUARD

| Order Group: | Platoon commander, sergeant, runner, two men with automatic weapons. |
| Three Fighting Squads: | Each comprising one NCO and seven men divided into a BAR (Browning automatic rifle) Squad and a Rifle Squad. Two men in each Rifle Squad also carried grenades. |

The other platoons in each company retained small arms but were reorganised on functional lines as a machine-gun, Spigot mortar or Smith gun platoon. Later the 1st Moray would organise a series of Junior Leader courses, training potential NCOs. These were courses conducted over four weekends and met the need of the unit for a steady supply of new leaders as existing NCOs dropped out due to age, work commitments or call-up to the forces. Other battalions adopted different structures, again reflecting the considerable degree of local autonomy enjoyed by the Home Guard.

Although Glasgow had developed an Area Town Fighting School this did not adequately meet the training needs of the 14 battalions in the city and local initiatives emerged to complement the Area provision. The 12th City of Glasgow Battalion, a works unit, created its own Town Fighting School at Garroway in the east of the city in a group of condemned houses and sheds owned by the engineering firm of William Beardmore & Co. and reconditioned by them for this purpose through the initiative of the commander of C Company, Major Snelus.

In 1943 Glasgow responded to the greater emphasis on battle training by establishing battle schools at High Craigton, near Milngavie, and at Darnley to the south of the city. Over 2,500 men passed through the High Craigton school. Craigton had been used as a range for grenade practice but in June 1943 the Sub District Commander, Brigadier Hobart, ordered its conversion to a battle school which was run by officers from the 1st City of Glasgow Battalion under Major Cowan Douglas. Over 1,200 men were trained at Darnley by officers and NCOs detached from the 3rd City of Glasgow Battalion. According to the historian of the 6th City of Glasgow (Corporation) Battalion, ". . .the men appreciated the necessity and usefulness of this training, and despite the rough going, little difficulty was experienced in obtaining volunteers to attend."[12]

This was a view that was echoed in the account of the 4th City of Glasgow Battalion, a works battalion based in the north-west of the city and made up of units from shipyards, engineering works and the BBC. Six battle platoons and 45 battle squads from this unit attended the High Craigton school: "The training received at this school was very popular in the Battalion, and every man who qualified there was proud to wear (unofficially) a purple distinguishing stripe on each shoulder strap of his battle dress. The Town Fighting School at Dixon's Blazes was also well attended. 60 officers and NCOs took the Course there, and were followed by large numbers of battle squads."[13] Many of the younger and fitter 4th Glasgow men who had gone through the High Craigton school were used in mobile platoons with the aim of providing reinforcements to other threatened factories and yards.[14]

In addition to the weekend and evening training offered at these schools many battalions organised regular weekend and also week-long summer camps. As the 5th Fife Battalion history said, "It was found that parading on Sunday afternoons or two nights per week that little real instruction could be given in Field Work so weekend camps were started."[15] The 5th Fife's camps ran from 2.30 p.m. on Saturday to 7 p.m. on Sunday and between 50 and 120 men attended each camp. This programme ran from May 1942 to September 1944 and the unit observed an increase in efficiency and an improvement in the health of the men. Camps such as these not only allowed more time for training but also had important morale benefits. The history of the 3rd Stirlingshire Battalion observes: "Living together permitted officers, NCOs and men to get to know one another better, with a resultant improvement in training."[16]

Many of these camps took place on farm land and the Home Guard often assisted with the work of the farm in any time left free from military training. "A Midlothian company, consisting largely of miners, repaid a farmer's kindness in this way just a week or two ago, and returned to their homes, at the end of a week's camp, with very pleasant recollections of their stay on the farm. Incidentally, these men had voluntarily sacrificed the whole of their annual holiday to attend camp."[17]

In Inverness-shire the 2nd Battalion organised what they described as weekend toughening courses at Fort Augustus. These focused on mobile warfare and included movement over difficult country by day and by night. The Home Guardsmen were out on these courses for 48 hours and might cover 25 miles during this time – a far cry from the image of the Home Guardsman as an old soldier, short of breath and with arthritic limbs. The 2nd Inverness-shire had also the great advantage of the commando training centre at Achnacarry being in their territory and the battalion became noticeably "commando-minded". Nor were they the only unit to emulate commando training – the 3rd Stirlingshire arranged a weekend camp as a commando course open only to those under 35 years of age. The 3rd City of Glasgow Battalion organised its men into four rifle companies, a machine-gun company and an additional company, Q Company; into this company were drafted all the youth of the battalion and their training was conducted on commando lines. F Company of the 8th City of Edinburgh Battalion had also been trained on commando lines and had 300 riflemen, 3 light machine-guns, 3 Browning automatic rifles and

4 armoured cars and was designed to form a striking force for Lothian Sub Area.

These voluntary residential training schools depended on men being prepared to give up free time or holidays to attend and in many cases they had to provide their own rations. A significant improvement came in 1942 when men sent on official training courses of at least one week's duration were given a loss-of-earnings payment of £3.14.0 per week. This greatly assisted training, especially in the case of junior leaders, who were potential section commanders. Some of the works unit were fortunate as employers often supported training.

A nine-day course in signalling was organised for Home Guards at a school in south-east Scotland organised by Scottish Command and attended by officers and non-commissioned officers who all had to be competent in Morse code before attending. Skills such as signals procedure, wireless and telegraph use, flag and lamp signalling and pigeon handling were taught. This was all quite a long way from the early days when the Home Guard was not encouraged to concern itself with signals and communications. The Scottish Command school was directed by a colonel of the Royal Corps of Signals.[18]

Valuable as such training camps were in imparting skills and building *esprit de corps*, much of the Home Guard's training inevitably had to be conducted on the regular weekday and Sunday parades. Typical of many training programmes was the 1st City of Edinburgh's recruit-training scheme based on two parades each of two hours' duration extending over a six-week period. The syllabus and time allocation were as follows:

Drill and handling of arms	3 hours
Elementary musketry, bayonet fighting and bombs	10 hours
Knowledge of company area and use of maps	2 hours

| Use of ground and Home Guard tactics | 5 hours |
| Anti-Gas and First Aid Training | 4 hours |

Such basic training had to be built on, skills honed and practical implementation of the core skills tested. This was done in advanced training, on firing ranges and in instructional classes at evening parades, but it took place above all at the weekends when more time was available and field exercises could be undertaken. The Home Guard had staged large-scale exercises early on it its history. The 3rd Stirlingshire Battalion, for example, ran a major exercise on Sunday 15 September 1940, when two of their companies acted as defenders of Grangemouth Aerodrome with the other two companies attacking the site. One issue thrown up by this exercise was the need for training of the umpires who determined the success or failure of the various operations.

The 6th City of Glasgow (Corporation) Battalion had one of its companies drawn from the staff of the Corporation Water Department and this company took on the task of guarding the Loch Katrine water pipeline, the main water supply for the city and its industries. Monthly exercises were held at various points along the pipeline, practising both attack and defence and gaining much experience in defending these vital points. To reduce the risk of air attack the pipeline was camouflaged by the engineer section of the 14th City of Glasgow (University) Battalion.

In 1943 Home Guard training was further formalised and standardised by the introduction of a proficiency test, with successful candidates wearing a badge on their uniform to mark their status. The 6th City of Glasgow Battalion managed to get around one-third of their men through this test in the year following its introduction.

In addition to recruit training and general skills training there was also the need for specialist training on heavy weapons such as the Smith gun, on signalling for signals sections, intelligence work for intelligence sections and for training potential officers. The 8th City of Edinburgh history, written late in 1944, notes that: "The greatest care has always been taken in the appointment of officers, and, their constant training <u>as officers</u> separately from other ranks and in addition to their other duties. During the past two years twenty new officers were required, and, only those who were selected for, and who went through the Battalion Officers' Courses and qualified were appointed."[19]

As we shall see when considering the Home Guard anti-aircraft batteries, these units made a practice of not commissioning officers in the battery, whatever previous military experience they might have, until they had passed out as qualified gunners.

Training was not confined to the men in the ranks and junior officers. Zone commanders from all parts of Britain were gathered at the Central Home Guard Training School in Surrey for a week-long course in July 1941.

When in March 1941 the names of the Scottish zone commanders – an appointment of full colonel rank – were announced, the list rather resembled an extract from *Who's Who*, featuring an Earl, two baronets, two knights and a number of individuals who had been at least a colonel in the regular army. Indeed, a month later the *Glasgow Herald*, a newspaper hardly known for its radical views, observed that of 319 names of senior commanders on the War Office list only 19 were plain "misters." Command selection boards were set up to handle commissions for ranks below that of battalion commander (lieutenant-colonel) and these boards included Home Guard representation. As an "authorised spokesman" told the *Scotsman*, "It is up to the Home Guard itself to see that it gets the officers it wants."[20] An interesting sidelight on the relationship between commissioned rank and class came in an instruction that men selected for the appointment of Platoon Commander, which in the commissioned rank structure would be held by a lieutenant, might, if he did not want to accept a commission, be granted the rank of Warrant Officer Class II (Company Sergeant Major.) This rank would allow him to command a platoon but presumably would relieve him from the burden of mixing socially with commissioned officers.

Not all appointments to commissioned rank went through smoothly. The non-commissioned officers and men of one west of Scotland company threatened to resign in December 1941 when a private was commissioned as a lieutenant, solely, they felt, on the basis of his social status. It was alleged that he was the least competent private in the company and had only attended two parades in the year.[21]

As was shown in Chapter 5, the Home Guard officer corps, if the evidence of one Borders battalion is to be relied upon, was relatively socially mixed compared with what one would have seen in an army officers' mess in 1939. Junior commissioned officers in the Home Guard could and did come from ordinary backgrounds even if the senior officers, the majors, colonels, etc. tended to be drawn from retired regular or Territorial Army officers, who in turn had tended to be drawn from the business, professional and landed classes. With the short life-span of the Home Guard it is hardly surprising that few men without previous military experience occupied posts higher than that of platoon commander. A four-year career was hardly long enough to rise from raw recruit to field officer. However there were instances, particularly in works units, where command seems to have gone to men with no military experience but high industrial occupational status; we shall later see an example where such an appointment proved extremely satisfactory. Nor, it must also be recognised, would it have made much sense to reject the substantial military experience that retired senior officers represented.

There was, however, a discernible democratic tendency. The distinguished academic Sir John Boyd Orr, recalling his experiences in the Aberdeenshire Home Guard wrote: "The units elected their own officer, and the best men chosen irrespective of their social standing. For example, Colonel Butchart, DSO, Secretary of the University, served in a group commanded by one of the janitors. Members of the aristocracy were in units commanded by a gardener or even a poacher."[22] This was undoubtedly true at section or platoon level but it was rather more common to find battalions such as the 1st Sutherland commanded by the Duke of Sutherland, the 1st Angus commanded by the Earl of Dalhousie, the 4th Inverness-shire commanded by the Viscount Gough and the 1st West Lothian by Lord Hope than to find janitors, gardeners or poachers at such command levels.

In many parts of Scotland Home Guard units took full advantage of being able to take part in exercises with regular army units. In most instances the Home Guard managed to give the army a hard contest and on many occasions the part-time volunteers got the better of the professionals. The 1st City of Edinburgh Battalion was tasked with the defence of Turnhouse Aerodrome and carried out regular exercises to test their defence arrangements for this important strategic asset. The battalion history tells the story of one such exercise in July 1942. A regular battalion of the Lowland Regiment reinforced by several hundred Home Guards together with armoured fighting vehicles attacked the battalion from south and east. After five hours' fighting in which the Lowland Regiment commander and HQ were captured and released three times, no enemy had succeeded in even reaching the RAF outer defences and the battalion had gained a signal victory. The history comments: "This was the most testing and realistic exercise the Battalion had. Unfortunately one Home Guard was killed during the action by a live round fired by the opposition."[23]

Private Charles Halloran was the unfortunate victim of friendly fire.[24]

Private Halloran's death is a salutary reminder that Home Guard training took its toll of men in death and injuries. Lieutenant William W. Jack of the 1st Lanarkshire Battalion, for example, died of the effects of gas fumes when he attended a course at the Glasgow Street Fighting School, while 2nd Lieutenant William Cook, an officer of 2nd City of Glasgow Home Guard battalion, was killed while supervising grenade practice by his men. "Lieut. Cook, who was 35 years of age, was educated at George Heriot's School, Edinburgh, and was employed as an engineer in Scotstoun. He is survived by his wife and two children."[25] As the *London Gazette* told the story when Lieutenant Cook's posthumous award of the King's Commendation for Brave Conduct was announced, he had been instructing his men in the use of No. 36 grenades with four-second fuses. "One grenade fell short, struck the parapet and fell at the feet of the thrower, who stooped to retrieve it. 2nd Lieutenant Cook, realising the danger of delay, dashed into the bay, pushed the man to safety, and himself seized the grenade but before he was able to throw it from the trench it exploded. He was mortally wounded and died three hours later. By his unhesitating action this very gallant officer, at the cost of his own life, undoubtedly saved that of one of his men."[26]

The 2nd Inverness-shire Battalion managed to conduct local exercises in its scattered area of western Inverness-shire around once a month and more extensive exercises involving regular troops training in the area three or four times a year. The 52nd Mountain Division and a Free French parachute battalion were amongst those who provided opponents for the Home Guard. In addition to providing opponents for mock battles the regular army units could also provide valuable assistance with weapons training, tactics and fieldcraft. Not all areas, however, enjoyed such assistance and the 1st Ross-shire Battalion found a marked difference between its east coast area where at various times the 51st Highland Division and Norwegian units were based and their west coast area which tended to lack regular units. When staff instructors were made available to the battalion their time was prioritised to training of the west coast sub-units. Other units found the availability of training from regulars was rather unpredictable; the Clydebank based 2nd Dunbartonshire Battalion had enjoyed a good relationship with 13th Battalion Argyll & Sutherland Highlanders, a home defence unit based in their area. However the Clydebank Blitz in March 1941 resulted in the Argylls moving away and the Home Guard had to rely on its own resources until in spring 1942 No. 12 Commando moved into the area and began to assist with training.

Training, as in the case of the 1st Edinburgh exercise mentioned above, was frequently focused on the operational commitments of a unit. These could range from the defence of essential communications facilities such as Post Office telecommunications centres, road and rail bridges, reservoirs, waterworks and airfields. In the case of the latter, it was necessary to liaise with the airfield operator – the Royal Air Force or the Fleet Air Arm – with the on-site defence force being responsible for the inner defence and the Home Guard unit for the wider defence and for the provision of a counter-attack force. A Fighter Command Control Centre was established in 1943 at Barnton Quarry in Edinburgh and was staffed almost entirely by members of the Women's Auxiliary Air Force, who were not authorised to carry weapons. The perimeter defence of this important strategic centre was entrusted to the 1st City of Edinburgh Battalion who redeployed one of their companies for this purpose. Later D Company from the 8th Edinburgh was attached to the battalion for protection duties in the quarry.

This arrangement seems to have worked well, but not every battalion was entirely happy with the role they were undertaking. The 9th Fife Battalion had amongst its commitments the important Royal Navy dockyard at Rosyth, a couple of naval stores depots at Crombie and Lathalmond and the north end of the Forth Rail Bridge. Most of the battalion's problems centred on Rosyth where the command of the base's defences was in the hands of the captain of HMS *Cochrane*, the RN shore establishment. He had a command post from which a land battle could not have been fought and in any case the dockyard defences lacked trenches and wiring. The writer of the unit history felt that a Royal Marine officer on the staff of HMS *Cochrane* would have been more attuned to land warfare and made a better defence commander.

There were in most battalion areas, apart from the vital points such as a road bridge or a reservoir, a limited number of so-called focal points – important towns, villages or cross-roads which needed to be held to the last man and the last bullet. Even though the Home Guard gradually took on a more mobile role and embraced the concept of offensive defence and battle drill, there still were static defence points; thus A Company of the 2nd Scottish Borders Battalion had six platoons based in Galashiels trained to hold the centre of the town. All roads could be blocked by steel tank obstacles, concrete rollers and barbed wire entanglements. A mobile platoon was held in reserve for the counter-attack role. The main role of the out-of-town platoons was the original Home Guard one of observation and harassing the enemy but in the event of a serious attack they would withdraw into Galashiels and aid the last-ditch defence. Many roads in this and other parts of Scotland were also furnished with fougasses or barrel flame traps – devices to pour flaming petroleum products over the road and so engulf enemy armoured vehicles in a sea of fire. In the rural part of Angus covered by the company-sized Tannadice Platoon of the 2nd Angus Battalion, the Finavon Static Platoon had six flame traps to assist in its defence of the Finavon Bridge, while the Glenogil Static Platoon defended Justinhaugh Bridge with four flame traps. Spigot mortars, a Northover projector and a Boys anti-tank rifle completed the heavy armament for these static defence detachments.

The 4th Dunbartonshire Battalion had four defended localities in its area: Cumbernauld and the Castlecary Viaduct, Kilsyth, Kirkintilloch and Auchenkiln crossroads. Its history notes with some pride that these defences were, "originally designed for defence from East but later made for all-round defence and were regarded as the finest defences in depth in the West of Scotland."[27]

The Castlecary Viaduct, carrying the main Edinburgh to Glasgow railway line over the main road from Glasgow to Stirling and Perth and the Forth and Clyde Canal was an obviously significant area and one which invaders, whether coming from east or west, would need to control to move across the country. In addition to the local defence provided by the 4th Dunbartonshire the Glasgow area commander had also assigned to the Castlecary Gap C Company of the 14th City of Glasgow (University) Battalion with the support of the battalion's artillery troop of four 75mm field guns. These guns, presumably of French origin, were replaced in October 1942 by eight 18/25-pounder field guns. The latter weapons were 18-pounder field guns that had been re-bored in the 1930s to fire a 25-pounder shell; by this time they had been replaced in Royal Artillery service by the purpose-built 25-pounder, this cascading of arms and equipment to the Home Guard being a very common feature. However not all Home Guard battalions operated field artillery and its allocation in this instance may have been due to the fact that 14th City of Glasgow was based on the

University Officer Training Corps which had the advantages of previous training and professional staff instructors.

The nature of the Home Guard's duties varied considerably over time. A very comprehensive account of these changes as they affected one rural battalion is given in the history of the 3rd Perthshire Battalion:

> The LDVF, though originally formed as an anti-parachutist force, quickly became road block minded and up to the end of 1940, the battalion's defences were centred round road blocks and road junctions, with Observation Points covering wide stretches of the area.
>
> The first major change came with issue of fougasse flame-traps to the battalion; these were widely installed and in order to utilise their possibilities to the full, the battalion trained ambush platoons and sections for operating them and guerrilla tactics. During 1941 this policy was developed and the battalion's role was to delay and harass the enemy without committing Home Guard forces to any "last man and last round" tactics and in the event of an enemy break-through in force, the battalion would retire to the hills and woods to continue in a guerrilla role.
>
> In 1942 the construction of Findo Gask RAF Station and the decision of Higher Authority to name Crieff as a focal point further changed the defence scheme; B Coy was made responsible for the defence of Crieff to the last man and round and elements of A & C Companies had to look to the defence of the aerodrome.[28]

These policy changes were at times driven by outside events as well as by the increasing efficiency and effectiveness of the Home Guard. For example, the capture of Crete in May 1941 by German airborne troops caused a major reappraisal of the value of static defences and a greater emphasis being given to mobile patrols and to what the 7th Fife Battalion history describes as, "some rather more systematic thought on the subject of the static defence of key positions, to which were assigned the older and less mobile members of the force."[29]

Speaking to journalists after a Home Guard exercise in Lanarkshire in April 1943, Lieutenant-General Sir Andrew Thorne, GOC Scotland, spoke of the need for the Home Guard to have a more offensive function rather than training solely for static defence and argued that the Home Guard should not waste time learning to use weapons which they would be unlikely to use in their area.

These changes must at times have seemed confusing to the Home Guard. The 1st Dundee Battalion had to work with three successive defensive schemes for the defence of their city The first of these, drawn up by the commander of Polish forces in the area, envisaged a defence of an outer perimeter roughly following the Kingsway – the orbital dual carriageway around Dundee built after the First World War – with an inner perimeter at key points. A later garrison commander planned a series of observation posts with manned roadblocks and a series of inner defended strong points, which even later was changed by a third commander to a plan basically designed to deal with the main perceived threat of the war's later phase – raiding or airborne parties – and which would concentrate on holding strategic points such as Balgay Hill, Dundee Law and key buildings in the city.

Mention of the Polish Garrison commander underlines the fact that the Home Guard was fully integrated into the command structure of Home Forces; although the Home Guard had its own command structure of area and group commanders, they also meshed into the regular army command, and would be operationally answerable to the local regular commander, sometimes with anomalous results. In March 1943 a Glasgow garrison was established with a regular lieutenant-colonel appointed as the city's garrison commander. He had under his orders no fewer than seven full colonels: the Home Guard zone commander, the zone second in command and five Home Guard group commanders, all of whom technically outranked him, a situation which could be found in many areas. The Ayrshire Zone history makes specific mention of the regular officer commanding Ayr Sub District, Lieutenant-Colonel J.C. Macindoe: "Under Colonel Macindoe we were a very happy family. He knew and understood the Home Guard, and his tact and experience overcame many a difficulty."[30]

The Ayrshire comment, in singling out one sub district commander, underlines the fact that not all such relationships were as harmonious. The complexities of handling large numbers of independently minded, self-willed, and at times prickly individuals must often have caused friction between regular officers and the officers of the Home Guard units assigned to their operational command. Although having Home Guard units operationally responsible to a regular officer was common, the reverse situation existed in some areas with a senior Home Guard officer being the garrison commander and regular units reporting to him. In August 1943 the officer commanding 1st Zetland Battalion was tasked with the defence of Lerwick, a function carried out by the battalion's E company with a mixture of Army, Royal Navy and Royal Air Force and a Norwegian commando company in reserve.[31]

After October 1941 the Home Guard branch at command HQ lost its executive role to the normal staff machinery and retained only an advisory and liaison role. While this probably helped the integration of the Home Guard into the national defence structure it had some unfortunate consequences. The writer of the Scottish Command overview of the Home Guard in 1944 observed with detectable weariness:

> Home Guard Liaison Officers found it constantly neces-
> sary to remind staff officers that a Home Guard Company
> or Battalion did not resemble even remotely a Company or
> a Battalion of any orthodox military formation. Considera-
> ble resentment was felt by many senior Home Guard com-
> manders at thoughtless demands or instructions received
> so frequently from Command and District headquarters.
> However as the Home Guard became more and more re-
> sponsible for the defence of the country, so did staff officers
> find it more and more necessary to become aware that the
> Home Guard had its being in different circumstances to
> those in which military courses and colleges of staff train-
> ing had decreed war could be waged.[32]

Some evidence of the demands that could be made from headquarters is seen in a somewhat tetchy response in February 1942 from Lieutenant-Colonel Burnett-Stuart, the command-ing officer of the 1st Aberdeenshire Battalion to a request from his zone commander for his defence scheme with plans show-ing battle positions and weapon strengths. He replied, not to his Home Guard zone commander but to the staff of Aberdeen Sub District who he clearly felt should have known the answers and should have asked him for any additional information that they did not have. He noted that Aberdeen and Fraserburgh had defence schemes drawn up by Aberdeen Sub Area but pointed out that he would need twice his current manpower to implement these schemes; he went on to say that the locations of all assembly points for the rest of the part of Aberdeenshire for which he was responsible, together with details of men and weapons, were already on file at Aberdeen Sub Area. Lieutenant-

Colonel Burnett-Stuart concluded with the somewhat barbed comment: "I regret that the facilities available at this Battalion Headquarters (the staff consists of one officer – the Adjutant & Quartermaster) do not admit of the production of detailed maps and plans in response to every request."[33] (In a previous life Lieutenant-Colonel Burnett-Stuart was General Sir John Burnett-Stuart, GCB, KBE, DSO etc, formerly Director of Military Operations and Intelligence at the War Office, General Officer Commanding British Troops in Egypt and prior to retirement General Officer Commanding Southern Command and an ADC to the king; he was presumably somebody who had a lot of experience of dealing with jumped-up junior staff officers.)

Burnett-Stuart's comment on the tiny staffs of Home Guard battalions was of course well founded. Home Guard battalions of over 1,500 men had from 1940 been allowed a full-time paid post of Adjutant/Quartermaster, initially recruited from within the Home Guard's own ranks but later filled from the regular army. In 1942, recognising that the workload involved in managing the administration and operations of a battalion which could number two or three thousand men could not efficiently be undertaken by one full-time officer and a clerical assistant or two, the War Office authorised the appointment of a quartermaster to each qualifying battalion. This officer handled the administrative side of the HQ work leaving the adjutant free to deal with training and operations, assisted by training officers from higher commands and sergeant instructors allocated from the regular army. Although as the war progressed Home Guard battalions developed quite elaborate battalion headquarters officer establishments, these were, with the exception of the adjutant and quartermaster, voluntary part-time posts. They could include an intelligence officer and a signals officer – as was the norm in an army infantry battalion – as well as a range of other posts such as an engineer officer, a bombing officer, gas officer, ammunition officer, liaison officer (responsible for relations with the local military), civil defence liaison officer, weapons training officer, public relations officer and a guide officer (in charge of the important role of guiding regular troops operating in the battalion area). Because the Home Guard was a part-time force many units found that they had to duplicate these specialist officers and NCOs; thus the 3rd Renfrewshire had both a musketry officer, responsible for training in rifle shooting, and an assistant musketry officer, who was the aptly named Lieutenant H. Carbine. In many cases these officers led a section or platoon of specialists in the various disciplines; for example, many battalions had an engineer platoon used to build roadblocks or construct and maintain firing ranges and bombing pits. In action, the intelligence officer and his section would, apart from gathering information on enemy forces, be responsible for security and for handling enemy prisoners. A point made in an army training memorandum about the problem of identifying the enemy was probably in the minds of many men in intelligence sections. The memorandum pointed out that German forces included many excellent linguists *au fait* with British slang: "It should therefore be impressed on all ranks that the use in conversation of "f-----s" and "b-----s" is *not* necessarily a guarantee of British nationality."[34] The 3rd Perthshire Battalion had undertaken the training of four or five men per company in intelligence duties during the first three winters of the Home Guard's life, but the effectiveness of this branch of service was handicapped by the battalion losing its intelligence officer in 1943 and being unable to find an appropriate replacement.

Traffic control duties devolved on the Home Guard in a number of areas and some battalions appointed a traffic control officer. This could be a burdensome but important job. When large army units were on the move, either for exercises or re-positioning, roads were likely to become congested; when one considers that an infantry division with an establishment of around 18,000 all ranks had around 3,300 vehicles and an armoured division had around 3,400 vehicles including 290 tanks the potential problem can be envisaged. Roads had to be available for both military and civilian purposes, and of course the roads of the 1940s tended to be narrower and mostly single carriageway; in rural areas they were often single-track with passing places and not tarmac surfaced. Moving even a single infantry battalion or a tank regiment through an area and keeping the roads open could be problematic. The small Kinross Independent Company, with a total enrolment of around 350 men, took on traffic control duties from April 1941 and found this a heavy responsibility due to the large number of main roads crossing its tiny county.

There was a tendency on the part of the Higher Command to look with some envy at the relatively manpower-rich Home Guard and seek to use it to solve its own problems. Such attempts to divert the Home Guard from what it saw as its core functions were frequently met with stiff resistance. In 1943 a request came from Perth sub district to the officer commanding the 6th Perthshire Battalion, based in the city of Perth, that members of the battalion be trained to operate a decontamination laundry in the city. This was presumably one of the plants of the well-known local laundry firm Pullars of Perth which had been taken over to decontaminate clothing after gas attack or anti-gas training. Lieutenant-Colonel Hunter, 6th Perthshire's CO, replied forcefully to this request. He made the point that to provide Home Guards for this task created the impression that all that had been said "by the Prime Minister and others as regards the national importance of the Home Guard is pure propaganda."[35] He went on so say that this proposal would lead to resentment among his men at being asked to carry out a task so completely different from the duties for which they had originally volunteered and that unfair advantage was being taken by the military authorities of the powers conferred on them to foist an unwelcome duty on the Home Guard. Hunter, who was a solicitor in Perth, had served in the First World War, rising to the rank of major and winning the Military Cross. He presumably felt, like his counterpart in Aberdeen, Lieutenant-Colonel Burnett-Stuart, that he need not defer too much to his military superiors and was not prepared to accept stupidity from Higher Command. He was evidently prepared to defend his men's interests against insensitivity and bungling in a way that might have been more difficult for a regular officer with a career to think about. Lieutenant-Colonel Hunter in a conciliatory close to his letter suggested that some men discharged from the Home Guard for unfitness or inability to attend parade might be used for this function.

One unusual task which came the way of the 12th City of Glasgow Battalion was to form and maintain from the personnel of three bank sub-units a bullion escort party to protect 80 tons of Bank of England gold which had been removed for safety from London to the Royal Bank of Scotland's chief Glasgow office in Royal Exchange Square. The Glasgow office had presumably been chosen for this function rather than the Edinburgh head office because of Glasgow's convenient proximity to the Atlantic sea-routes. The idea of evacuating the nation's financial reserves overseas in the event of an invasion – probably to Canada – was evidently being planned for and the bank staff were thought the most appropriate people to deal with this task. Interestingly enough, the Bank of England Foreign Exchange Control staff had been relocated from London to Clydeside early in the war. Personnel, arms, equipment and rations were detailed and maintained for muster on receipt of a special code word from August 1942 to October 1944.

Less contentious than running laundries was the use of Home Guardsmen for tasks such as mine-watching or bomb disposal. The threat of isolated German bombers dropping mines in the congested waters of the Clyde was a very real one, with significant implications for what was a critically important and extremely busy port and shipbuilding centre. Day and night on the Clyde, the 4th City of Glasgow Battalion, a works battalion drawn from factories and shipyards in the north-west sector of the city, manned 11 mine-watching posts on the river. In order to provide a sure fix on any mine it was the practice to try to obtain at least two sightings from separate observation points. Because of the winding nature of the river and the obstructed view caused by docks, shipyards, and indeed ships, an extensive network of spotting points had to be maintained. These mine-watching duties also involved personnel from other battalions, including the 5th Lanarkshire, who manned a mine-watching point at Yoker.

This reflects the way in which geographical boundaries could be quite flexible. The 5th Lanarkshire was a unit based in the Rutherglen/Cambuslang area, quite distant from the extreme western boundary of Glasgow at Yoker. However it incorporated a number of works sub-units, including one, F Company, based on the Clyde Valley Electrical Power Company which had detachments at Lanark, Motherwell and Yoker Power Station. It was considered better for the detachment at Yoker to report to the distant F Company commander than to the geographically closer but organisationally separate 4th Glasgow Battalion. The link between employees of an organisation was thought to outweigh, at least for organisational and training purposes, the tidier geographical links to a local unit. Thus the 8th City of Glasgow (2nd LNER) Battalion had railwaymen working between Corrour and Mallaig administered from its Glasgow headquarters, and the 10th City of Edinburgh (3rd LNER) Battalion had a company at

Hawick and an independent platoon at Eyemouth. Similarly the three Post Office battalions in Glasgow were grouped together in a Post Office Home Guard District rather than being administered by the geographically appropriate sector within Glasgow Zone. Although these arrangements prevailed for administration and training, in the event of active service they would liaise with other units in their geographical area.

Auxiliary bomb disposal units were created in 1942. Glasgow had six units with a total strength of 142 men formed within three of the works battalions and based in factories such as North British Loco (12th Battalion) and Albion Motors (4th Battalion.) Other industrial areas also produced small trained bomb disposal teams. The 2nd Lanarkshire had a 50-strong squad based at Stewarts & Lloyds Steelworks, Coatbridge. Some of these men were sent to England to assist in the aftermath of heavy bombing raids. Assistance was given to Civil Defence by a number of battalions creating fire-fighting squads, demolition squads and decontamination squads; these specialist squads were often provided by works units or sub-units.

During the Clydeside air raids of 1941 many Home Guard units and men were involved in rescue and demolition work and their contribution was recognised by a number of awards. Lance-Corporal A. McLean of the 1st Renfrewshire Battalion was given a certificate on parchment for gallantry and devotion to duty during air raids on Greenock. He had been in charge of a small group of volunteers and remained continuously on duty during the second night of the raid and the following day. He had helped to dig people out of ruins and shepherd them to safety, rendering great service and showing fine initiative. In civil life Lance-Corporal McLean was a draughtsman. Another Renfrewshire Home Guardsman, Private Donald M.G. McKay, lost his life while assisting Civil Defence in Glasgow. He served in the 3rd Renfrewshire Battalion and the unit history suggests that his poor eyesight, resulting from First World War service, contributed to his fall from the upper floor of a building where he was attempting a rescue.

In May 1941 the Nobel Explosives factory at Ardeer, near Irvine, was bombed and many grass fires started in the complex. The inlying piquet, or watch, of the 8th Ayrshire Battalion was released from defence duties for fire-fighting duties. It also mounted guard at the main gate to prevent unauthorised personnel entering and blocked roads within the factory complex where unexploded bombs were located.

As a part-time force, many of whose members were in key jobs in civilian life, there were concerns that if an emergency arose and the Home Guard was mustered – that is, called to full-time service – a great deal of disruption to essential services and functions would ensue. There was, for example, an issue with regard to teachers; there would be a problem if children had to be evacuated from an area and the teachers assigned to escort them were already engaged in full-time Home Guard duties. Similarly, railwaymen would be needed to keep the trains running, but this could conflict with their defence duties. Eventually in February 1942 a solution was found whereby men in the Home Guard were divided into two groups: those who would be immediately available in the event of an invasion, and those who would continue in their civilian roles and not be mobilised until fighting was expected in their area.

Works battalions formed a very significant part of the Home Guard; in Glasgow, for example, more than half the 20,000 members of the city's 14 battalions were in works units that needed to be involved in the overall defence of the city. The effectiveness of a works unit depended to a considerable degree on the co-operation offered by the employer. Colonel A.K. Reid,

Home Guard adviser to Glasgow Sub-District, wrote: "In works units individual training was usually done in accommodation put at the disposal of the Home Guard by the managements of the various firms and undertakings concerned, all of them taking a very keen interest in their contingents. This applied with particular force to the two railway companies, which provided and furnished guard rooms and miniature ranges and generally gave their units the greatest encouragement."[36]

Although the Post Office and railway battalions were officered by the management of these institutions, most works units found that outside senior officers were advisable. The history of the 12th City of Glasgow Battalion wrote of the: "tendency in each works to think in terms of its own Sub Unit individuality rather than in terms of being part of a Company or even Battalion. Partly to counter this tendency, and partly to save other possible embarrassments, the principle was established that HQ Staff Officers, Company Commanders and second in command of companies would not be drawn from personnel who had any connection with any of the Works attached to the Battalion."[37]

The account of the 6th Lanarkshire Battalion, which combined a mobile company, a railway company and 18 works sub-units of company or platoon strength gives an interesting insight into the problems of works units. The writer, Lieutenant-Colonel W.D. Macrae, notes that the appointment of company commanders had taken place without friction despite being a matter of potential concern.

> To the ordinary regimental officer the foregoing might well seem superfluous but when a Home Guard Commanding Officer looks back over these years and thinks of the time and anxious thought expended, the constant tact required for the composing of minor differences and the ironing out

of difficulties, and the rivalry between the works which had to be welded together he might feel qualified in post war years to practice the noble art of Isaac Walton which requires above all other qualities that of infinite patience. He did indeed fish in troubled waters. He was supplied with a line but had no rod for chastisement.[38]

The same account gives an example of one officer, Major James Cunningham, who was managing director of the Etna Iron and Steel Company, and despite having no previous military experience commanded the 6th Lanarkshire's B Company with such success that in August 1944 he was appointed as second in command of the Battalion.

Communications problems were referred to in Chapter 5 and the initiatives taken by some units, despite official lack of enthusiasm, to create a dedicated signals service. An older form of communications was also discovered to be of value in modern warfare. Homing pigeons were found by many units to be a useful ancillary means of communication. Lieutenant G.G. Johnstone furnished the 1st Ross-shire Battalion with a pigeon service using birds from his own loft and, when all other means of communication failed, the pigeons got through. The 1st Moray also had a pigeon-post service for a time but discontinued it, in part due to the scarcity of pigeon food. Glasgow Zone had a particularly well-developed pigeon service, reflecting the popularity of pigeon fancying in and around the city, and it drew on the generous co-operation of civilian owners. There was a pigeon loft at zone headquarters with lofts at various battalion headquarters and on the line of the vital Loch Katrine pipeline. A pigeon officer, Lieut. T.H. Rennie, was appointed to zone staff and battalion pigeon officers were authorised for the 1st, 2nd, 3rd, 6th and 11th City of Glasgow Battalions. In Shetland carrier pigeons were held by

D Company (Walls) and F Company (South Shetland) and flew successfully between these locations and headquarters.

The Home Guard seems to have been able to call on a great wealth of ingenuity to improvise solutions to problems. The creation of home-made grenades and anti-tank mines was discussed in Chapter 5. However, ingenious devices were not just confined to explosives. The 3rd Dunbartonshire Battalion had to man important defensive positions on the perimeter of their area, at Auchineden and Strathblane. The unit's history notes: "To enable platoons living in Bearsden and Milngavie to man positions at Auchineden and Strathblane with any degree of speed, it was deemed necessary to devise means of making machine-guns mobile. This was effected by mounting most of these weapons on motor-car trailers contrived from second-hand motor-car wheels and axles. These trailers were noted and copied by other Home Guard units even in the South of England."[39]

Characteristic of the Home Guard was the use of whatever local resources were available. When the 4th Perthshire Battalion wanted to train their men in the art of interrogating suspects they turned to a non-military source, the Perth Repertory Theatre, for help. Four actors from the theatre played the roles of German spies and refugees, telling convincing tales of misfortune. "All four entered very thoroughly into their job and were not upset by a certain amount of pretty severe questioning, and were quite willing to be searched. On their own they invented some very ingenious hiding places for messages, etc."[40] Other civilian specialists were used to assist in Home Guard training: scoutmasters to teach map-reading and fieldcraft, architects and builders to show how to adapt houses to a defensive state, artists to instruct in camouflage.

The novelist Compton Mackenzie, commanding the Barra Company of 2nd Inverness-shire Battalion, found a particular problem with one specialist. With a small and largely elderly population, most of the younger men being in the navy or the mercantile marine, there was only one man on the island with the technical skills needed to keep vehicles and machinery in operation. As such he was clearly indispensable both to the normal life of the island and to the efficient operation of the Home Guard. However, he received his calling-up papers for the Royal Air Force, and despite a lengthy campaign by Mackenzie was eventually obliged to leave the island. Local loyalties and local circumstances in places like Barra could greatly affect the operation of the Home Guard. When all the Outer Hebrides units were detached from their mainland parent battalions in Inverness-shire and Ross-shire and the 1st Hebrides Battalion was formed the Barra Company became a platoon. Mackenzie, who was in poor health and had work that regularly took him off the island, felt that it was an opportune moment to resign. However his second in command, the local parish minister, was expected to be called up for service as a naval chaplain and there seemed to be an absence of men able to assume the role of sergeant. There was also a great antipathy among the Barra men to being commanded by men from the neighbouring island of South Uist. Mackenzie wrote to the commander of the 1st Hebrides Battalion: "The men here as you know have never been soldiers, and though it is easy enough for me to manage them, and they have given me excellent support, they do not like the idea of being a platoon attached to, and as they think, under the command of South Uist. All very silly no doubt, but I don't have to explain insular prejudices to you."[41] In the event Mackenzie was persuaded to remain in command of the Barra platoon.

The one resource that the Home Guard was never able fully to call on was the nation's women. From the earliest days there had been women assisting with typing and other clerical work and

the campaign of Dr Edith Sumerskill MP to allow women into
the Home Guard on the same basis as men continued. Despite
official hostility Dr Summerskill had set up her own unofficial
group, Women's Home Defence, which had 20,000 members in
200 units by July 1942. Eventually in April 1943 a small concession was made and Sir James Grigg, Secretary of State for War,
told the House of Commons that women were now entitled to be
nominated to serve as auxiliaries in the Home Guard. However
no uniforms would be issued and their duties would be confined
to non-combatant duties such as driving, cooking, cleaning and
clerical work. There was certainly to be no question of arming
these women or training them in firearms. A plastic badge, which
did not require polishing, would be issued as the sole sign of service.

Like many concessions by a reluctant government to popular
opinion this half-hearted response caused as many problems as it
solved. The 1st Fife Battalion's history sums up the experience of
many units. It observed that throughout the unit's existence considerable assistance had been given by women in clerical work at
battalion and company headquarters. The decision in 1943 however raised issues: "They were to be called Nominated Women
but this name was considered most inappropriate and caused
such a furore that it was finally changed to Home Guard Women
Auxiliaries."[42]

There were never large numbers of these women auxiliaries;
Fife and Kinross Zone had some 13,000 men of all ranks on its
books at stand-down, and an establishment for 428 auxiliaries.
The rules about what women could and could not do were very
clear, and were also applied to services like the Women's Royal
Army Corps and the Women's Auxiliary Air Force; they unam-
biguously prohibited the bearing of weapons, or training women
in the use of weapons. However it is also clear that, as far as the

Home Guard was concerned, these rules were being ignored. In September 1942 a Home Guard Directorate conference was told that the order of the Commander in Chief Home Forces that women were not to be trained in the use of arms was being ignored and the Command representatives present were told that any breach of this order would be dealt with as a disciplinary offence.[43] This threat of sanctions, which was widely disseminated, would hardly have been necessary if there had not been ample evidence of Home Guard units training women in handling arms.

If women could not be trained to bear arms there was less of a problem with young people. Although there was a minimum age limit for the Home Guard of 17 it was realised in 1941 that there was a declining pool of older recruits for the force as military conscription extended its reach. It was therefore decided to establish formal links between local Home Guard units and the cadet force with the intention that the cadet forces would serve as a feeder to the Home Guard. The 2nd City of Edinburgh Battalion organised a recruiting week in January 1942 for the cadet company which was affiliated to it. Pre-war financial restrictions and loss of drill halls had reduced what had once been a flourishing army cadet force in the city to something much less extensive and the local cadet movement had to be rebuilt, to a considerable degree. The rise of the Air Training Corps had proved attractive to many teenage boys and posed a significant challenge to the cadets and the Home Guard for the service of the available youngsters.

No Scottish military formation ever seems capable of existing without immediately creating a pipe band and the Home Guard was no exception. There was, however, no question of official sanction or support for pipe bands; there was, after all, a war on, and scarce resources could not be diverted to such fripperies. As late as July 1943 Arthur Woodburn, Labour MP for Clackmannan and East Stirling, asked if kilts could be issued to Home Guard bands and was advised that the Home Guard was not authorised to have pipe bands and so kilts could not and would not be issued. In 1944 the Secretary of State for War refused to let Home Guard units purchase War Office kilts.

Despite this there were large numbers of Home Guard kilted bands, although one or two Glasgow battalions had to have their pipes and drums in battledress. A variety of approaches were used to ensure that the Home Guard could march to the sound of the pipes. A number of battalions had bought kilts from private contractors before clothing rationing was introduced in June 1941, and in one case the entire membership of a civilian pipe band, the Polmadie LMS Railway band, joined the 7th City of Glasgow (LMS) Battalion and brought their pipes, drums and clothing with them. Nor was the formation of bands solely a battalion-level concern; B and D Companies of the 3rd Renfrewshire Battalion each raised and maintained pipe bands from their own local funds and the 3rd Dunbartonshire had both a battalion pipe band and a band in S Company which recruited in the Killearn, Drymen and Strathblane area of Stirlingshire. The Callander detachment of the 5th Perthshire Battalion could muster three pipers, sadly in battledress rather than in kilts.

Some creative thinking went into equipping Home Guard pipe bands. The 4th City of Glasgow Battalion maintained no fewer than five bands. Most of the instruments came from previously existing works bands, while kilts were bought just before the introduction of clothes rationing; sporrans were obtained through the good offices of the Territorial Army and hose tops, flashes and ankle puttees were issued in error by the army ordnance department and not returned! The quality of Home Guard bands was often high. The 2nd Lanarkshire Battalion's band won the Scottish Wartime Championship and broadcast on BBC Scotland in June 1943.

Nor were pipe bands the only form of military music found in the Scottish Home Guard. The 11th City of Glasgow Battalion, not content simply with forming a pipe band in September 1940, formed a brass band at the same time, which it claimed was the first Home Guard brass band in Britain.

Pipe bands may not have been officially sanctioned but they played an important part in many official Home Guard events. In February 1941, Lieutenant-General Sir Robert Carrington, GOC Scotland, inspected what was described as the largest Home Guard parade to date in the West of Scotland in Glasgow. Five Home Guard pipe bands appeared before him, without apparently incurring official displeasure. It became customary to hold anniversary parades in May to mark the foundation of the Home Guard. The first of these parades took place in 1942 and such events would have been incomplete without the pipes and drums to lead the parade. The Polmadie Band of 7th City of Glasgow Battalion was often called on to travel down to Gourock or Greenock to give an authentic Scottish welcome to troops arriving at the Tail of the Bank in troop convoys or on giant liners like the *Queen Mary* and the *Queen Elizabeth* from Australia, Canada, New Zealand or the United States. In March 1941 King George VI and Queen Elizabeth visited Rosyth Dockyard, where the King inspected the 9th Fife Battalion, raised from the dockyard workforce, and the unit's pipe band played the "Skye Boat Song" as a royal salute.[44]

Just as marked as the Scottish military fixation with the pipes was the urge for groups to break into print. Paper was scarce and rationed during the war with printers desperately trying to lay their hands on supplies to meet the demand for official publications, recreational reading and the countless forms that a burgeoning bureaucracy demanded. Despite such restrictions, units such as the 2nd City of Edinburgh Battalion and the 2nd

Lanarkshire Battalion managed to produce battalion magazines and others published unit histories at the end of the war.

Life for the Home Guard was seldom static; no sooner had some state of satisfactory equilibrium been reached when higher command intervened and forced a radical change on the organisation. Perhaps the most extreme form of this was the transfer of large numbers of Guardsmen into anti-aircraft duties, a development that will be discussed in a later chapter. Less obvious perhaps were the consequences; late in 1941, when the age limits for military conscription were raised to include men between 41 and 51, this had the potential to remove from the Home Guard many of its fittest and ablest men, who were also often men who had served in the First World War and could bring valuable combat experience.

At the same time there was a growing feeling that there was an inequity in the fact that many men were spending long hours training to defend their community through the Home Guard while others, equally able to serve, remained quite uninvolved in the national struggle. This point was impressed on the government by the Earl of Mansfield who told the story of six farmers in his area of Perthshire, four of whom joined the Home Guard and two did not. Two of those who joined later resigned: "But the two others, whose circumstances are precisely similar to the other four, continue to come in after their day's work for a distance of several miles over very rough roads when they have been working for many hours at their ordinary occupations. They are regular attenders and are among the most satisfactory men I have got. Therefore, I ask, is it fair that the other fellows should escape simply because they do not like to do a lot of extra work on the top of what they are already doing?"[45]

There was also, in some areas, a shortage of volunteers for the Home Guard, at a time when the invasion threat, though diminished, had not vanished. The German invasion of the Soviet Union had ground to a halt, and with the advent of winter looked unlikely to produce a swift victory. However if Hitler was unwilling to commit to a long and bloody campaign rather than the lightning victories which he had achieved to date, this might, it was thought, lead to the disengagement of German forces in Russia. This would allow their subsequent redeployment to the west with a seriously increased risk of raids on Britain or a full-scale invasion. Large elements of the army were fighting overseas and home defence would have to rely to an increasing degree on the contribution the Home Guard could make.

In March 1942 the Earl of Elgin, Zone Commander for the Home Guard in Fife and Kincardine, spoke on the introduction of compulsion to the Home Guard. In the debate in the House of Lords, initiated by the Earl of Mansfield, he cited the example of a Fife battalion that had shrunk from 2,000 men to 1,254. Its commanding officer had said that it was his opinion and that of his company commanders that: "the men are available in this locality, which is wholly a mining community, and hundreds of young men are therefore reserved and should be compelled to give service either in the Home Guard or Civil Defence Service. The need for weapons does not appear to be of so great urgency as the fact that the men cannot at present be trained to use these weapons."[46] Lord Elgin went on to say that the professional opinion of the senior officers of this battalion was that the manpower needed to carry out the duties allocated to them was 3,400, or about 2,100 men more than they currently had. The men, he emphasised, were available, and all that was stopping their utilisation was the absence of compulsion.

The government initially decided to bring compulsion into the Home Guard in just some areas of England, but eventually it was introduced nationwide. The measure provided that men between

18 and 50 who were not otherwise engaged in some form of Civil Defence service could be directed by the Ministry of Labour and National Service into the Home Guard. With this came a legal obligation to attend drills and training and the provision that failure to do so would lay the guardsman open to prosecution in the civil courts rather than through courts martial. The original LDV scheme had been carried over into the Home Guard and men had enjoyed the right to resign from the service on 14 days' notice. This was obviously incompatible with the new compulsory service regime, and the right to resign, what had been often jocularly referred to as the "housemaid's clause," was withdrawn, although existing voluntarily enrolled guardsmen were given a short period in which they could resign from the Home Guard. Most of those who took advantage of this were older men. However some younger men resigned and were happy to be directed back into the Home Guard because their "drafted" status made attendance at drills and parades a non-voluntary matter; it was thus easier to justify to employers, family and churches, all of whom might have had a degree of doubt about the validity of their reasons for absence from work, family commitments or worship. One of the changes that accompanied the introduction of compulsion was the last step in the militarisation of the Home Guard; commissioned and non-commissioned ranks had been introduced earlier and now volunteers would become officially known as privates. Another consequence of these changes was that the officially authorised maximum attendance at drills and training was established as 48 hours in a 4-week period. Men could voluntarily do more than 48 hours but could not be made to do so and, of course, such limits were abandoned if and when the Home Guard was to be put on an active service footing.

Much emphasis was placed on the need to retain the volunteer ethos. A *Times* leader said: "Finally, it cannot be too strongly insisted that it is the intention of the War Office to preserve the character and spirit of the force and the general atmosphere which permeates it and which is largely due to its voluntary nature. Every care will be taken to ensure that commanders realise the extent of their latitude as regards local needs."[47]

Colonel Colville was MP for Midlothian and Peebles and the senior staff officer in Scottish Command dealing with Home Guard matters. He had been a Home Guard area commander and spoke in the Commons on compulsion in December 1941. Acknowledging that compulsion was an essential development, he pointed out that judgement and experience would be needed in introducing it if the needs of industry and agriculture were to be safeguarded and if the "great and fine spirit which has animated the vast majority of members of the Home Guard for one and a half years is to be fully maintained."[48] Colville went on to argue that compulsion would not only give the Home Guard the manpower it needed for its ever-widening duties, but it would be more equitable and would deal with the small minority of existing home guardsmen "not pulling its weight in duties or in training". With the advent of compulsion the Home Guard by July 1942 could plan for a UK strength of over two million men, a far cry from Eden's original hope for 200,000 men.

Although the compulsory direction of men into the Home Guard had been initially introduced in limited areas this was seen as potentially divisive, with compulsion in force in one county but not in the next. In March 1942 Duncan Sandys, the Financial Secretary to the War Office, advised the House of Commons that compulsion would be introduced progressively throughout the country and that this measure would not only meet operational needs but would let the country at large see the importance that the government placed on the role the Home Guard played in national defence.

There was naturally concern within the Home Guard about the implications of the introduction of compulsion and fears that its character might change. Fewer men than might have been expected took advantage of the opportunity to leave. The 6th Lanarkshire Battalion, which drew heavily on the iron and steel trades of industrial Lanarkshire for its men did lose a large number of men at this point; its strength sank from around 3,000 to 1,900 but special circumstances applied here. Many older men who were working long hours in these heavy industries found the military commitment to be a strain on their health and took the opportunity to retire from the Home Guard with honour. Concerns that the new "directed men" might not be of the same standard as the volunteers soon proved unfounded. The 6th City of Glasgow Battalion's experience was that: "Contrary to expectation the directed men turned out to be of first class quality and soon made themselves at home in their new surroundings."[49]

The original voluntary nature of the LDV and then the Home Guard had meant that discipline was not much of a problem; all the members were volunteers, and although they might initially occupy different appointments or ranks after they were introduced, there was a common purpose and a common spirit. Leaders or officers had to lead by example and character rather than by adherence to a rulebook or army regulations. If a home guardsman did not like the way an officer was treating him or had no confidence in him or got fed up turning out for parades on wet Tuesday nights, then, in the period before compulsion, he could simply stop attending or formally resign. Compulsion removed that option and brought into the force a small percentage of men who did not want to be there and were unwilling to pull their weight. There needed to be some sanction to ensure that men received a minimum amount of training and attended a minimum number of parades. The War Office was still reluctant to abandon the voluntary principle or to have the Home Guard treated as a normal military force fully subject to court martial law and the king's regulations at least as regards attendance, although offences committed when on duty were dealt with by military law. Clearly with an unpaid force the normal minor military sanctions of additional duties, restriction of leave and stoppage of pay were hardly viable. Failure to attend parades and training was thus dealt with by the criminal justice system; thus, for example, four Home Guardsmen in Clackmannanshire were fined £15 each with the option of 30 days' imprisonment for missing parades. Two said in their defence that they simply did not want to go to parades, one of them claiming he had joined the Home Guard as a dispatch rider but his bike had broken down and he was transferred to infantry duties which he did not want. The third of this batch of offenders said he was tired coming from work and had gone to bed while the fourth told the court that he had been on back shift for a time. The 7th City of Glasgow (LMS Railway) Battalion reported that there had been no need for enforcement of discipline before compulsion, but afterwards fifteen cases of failure to attend parades had been processed and two cases of contravention of the Army Act by NCOs acting under the influence of drink were recorded. One of the Army Act offences was dealt with by admonishment by the sub-district commander and the other by reduction of the offender to the ranks.

There were relatively few prosecutions for absence from parades because there was a definite policy only to prosecute when there was a clear case. Lord Croft, Under Secretary of State for War, explained the official policy in a debate in the House of Lords: "I suggest to your Lordships that it is much better to take great care over such prosecutions and to secure a high percentage of convictions rather than to adopt less strict methods which would lead to a number of acquittals, so largely defeating the purpose of the prosecution in the minds of the public."[50] He went on to state that of the 202 prosecutions which had been authorised, 89 per cent had resulted in a conviction, in 5 per cent the police had decided not to prosecute, in 1.5 per cent the Home Guardsman had been admonished and only in 3.5 per cent of cases was the accused acquitted. From Lord Croft's comments it is evident that the chief object of prosecution was not the punishment of the guilty few but the deterrence of the many.

Home Guardsmen were subject to courts martial under the normal army code. There was a case in Edinburgh, for example, where a sergeant in the 3rd City of Edinburgh Battalion who was on active duty as the non-commissioned officer in charge of a guard had neglected to ensure that the guardroom was never left unattended. He was found guilty of conduct prejudicial to military discipline and good order and demoted to corporal. While in the regular army this would have entailed a financial penalty, in the unpaid Home Guard it simply represented a loss of status.[51]

The 1st Stirling Battalion's experience of compulsion was that: "Direction at first produced a lot of first class material but it was soon found that there wasn't an inexhaustible supply and many men employed in mining, railway, ordnance etc found it impossible to give to the Home Guard the time needed for efficient training."[52] This latter point was of course one that applied across the country and affected many Home Guardsmen who were juggling jobs (often with wartime overtime added on), family commitments and Home Guard duties.

The case of an East Lothian miner who was fined £3 in September 1942 for refusing to do night duty illustrates the point. The National Union of Mineworkers took up the issue and argued that a man could not be expected to do his next day's shift in the mine if, after a day's work, he was expected to do night duty in the Home Guard. The interim president of the Scottish NUM, William Pearson, said: "We have no objection to Home Guard duty up to a reasonable hour of night – say, between 5 p.m. and 10 p.m., but we do object to rest being disturbed. It is a physical impossibility for a man to carry on like that."[53]

Major A.J. Mornard, the commander of E Company 7th City of Edinburgh Battalion, who had initiated the prosecution of the fined miner, later wrote to the *Scotsman* to point out that the night duty complained of was in fact night guard duty and simply required the man to sleep at the company headquarters on a rota of one night in three weeks. A miner on day shift worked from 6 a.m. to 2 p.m. and was not required to report for guard duty until 10.30 p.m. Major Mornard pointed out that the night guard duty had two functions: to stop intruders entering the premises and to operate phones in case of emergencies. He wrote that it was not considered necessary for either of these functions for the guard to stay awake: "burglars, fifth columnists, and the telephone will all make sufficient noise to wake them."[54] Major Mornard admitted that sleeping at company headquarters (the Wallyford Miners' Welfare Institute) was not quite the same as sleeping at home in one's own bed. He pointed out, however, that most of his company were in fact miners on day shift and had been doing this night guard duty for over a year without problems. He concluded by pointing out that other miners worked just as hard as the man who had refused night duty and it was unfair on them if he was allowed to refuse to take his share of the work.

In August 1942 a factory liaison officer was appointed in each command to advise on problems arising from Home Guard duties and employment. It was emphasised that home guardsmen in defence production should not be asked to undertake duties that would impair their work efficiency. In February 1943 the Trades Union Congress made representations to the War Office about the situation of Home Guardsmen working long hours in civilian employment and suggested that once they had reached a satisfactory level of military efficiency they might be allowed to ease off training. The War Office responded to this by advising that defence production workers and agricultural workers should have calls on their time for Home Guard training reduced as far as possible.

Many Home Guard units in rural areas recognised that the age-old seasonal demands of agriculture had to take priority over military training, and in any case the country desperately needed all the food it could grow or raise to reduce its dependence on foreign imports, which were brought into the country in the teeth of the German U-boat menace. The 3rd Scottish Borders Battalion, based in Berwickshire, normally paraded three Sundays out of four and on two evenings a week shut down for training purposes during the harvest season: "In a county where 90 per cent of the male population is agriculturally employed it could not be otherwise, and the break prevented boredom."[55] Likewise, in the grain-growing country of Morayshire the 1st Moray Battalion managed to reduce drills and training in 1942–44 to allow men to work on the harvest. Similar arrangements were made to release shepherds at lambing time.

If the demands of crops and animals could conflict with Home Guard training so too could the demands of the church in a period that was rather more churchgoing than ours. The complaint made to Deer Presbytery in 1940 was discussed in Chapter 5 but the issue returned more generally, especially when the invasion threat was seen to have receded. In a period when most people worked a 5½-day week, Sunday was the only full day free for Home Guard training. The Rev. Matthew Stewart, Convener of the Church of Scotland's Church and Nation Committee, speaking at the 1942 General Assembly, condemned regular Sunday drills and parades and noted that many areas managed to have Home Guardsmen reporting for duty on alternate Sunday afternoons. He acknowledged the national importance of the Home Guard and felt that no one would care to see any steps taken to impair its efficiency but felt that it must not deprive its members of the chance to take part in Sunday worship.

The question of a conflict between the Home Guard and religious observance was solved in a variety of ways across Scotland. In Glasgow the 6th City of Glasgow (Corporation) Battalion appointed the Rev. A. Neville Davidson, Minister of the Cathedral, as padre to the Battalion and annual church parades were held from 1942 to 1944. The 1st City of Edinburgh Battalion had as their padre the Rev. R.W. Leckie of Davidson's Mains Church and the battalion's last corporate act was to hold a service at Cramond Kirk. E Company, the Melrose Company, of the 2nd Scottish Borders Battalion got around the problem of Sunday parades by organising a short voluntary religious service in the Drill Hall before the parade each Sunday. These services were conducted on a rota basis by the various ministers of the town and were very well attended. A War Office edict in May 1941 had ruled that no commissions as chaplains to the Home Guard would be allowed but that battalions and sub-units could make honorary appointments. These honorary padres would not wear uniform unless they were enrolled members of the Home Guard. Not all Home Guard church parades went off entirely smoothly. Bill McChlery recalls a parade organised for the 4th Glasgow Battalion in the Ascot Cinema at Anniesland Cross, Glasgow. Unfortunately nobody had thought to remind the men that a church parade was a "no-smoking" event and during prayers around 200 men lit cigarettes for a crafty smoke – to the horror of Bill McChlery, a regular church-goer and sergeant in the Boys' Brigade.[56]

An interesting sidelight on this issue was that in 1941 the

Lord High Commissioner to the General Assembly, the Sovereign's personal representative to the Church of Scotland's principal deliberative body, Sir Iain Colquhoun of Luss, had requested that his guard of honour be provided by the Home Guard; a detachment of the 3rd City of Edinburgh Battalion under Captain A. Kay, MC, filled this role. Colquhoun's successor as Lord High Commissioner in 1932 and 1943, the Marquess of Montrose, continued the custom of honouring the Home Guard in this way.

If the spiritual welfare of the Home Guardsmen had to be attended to, so too did their medical care. When the LDV was established little thought seems to have been given to this matter and the initiative in fact came from the medical profession, namely the British Medical Association's Central Medical War Committee, which offered the provision of local medical and first aid services to LDV members injured on duty. It must be remembered that this was before the establishment of the National Health Service and that army medical facilities were not going to be able to cope with the LDV on top of an already expanded army. In August 1940 the War Office communicated the BMA's offer to commands and requested that full advantage be taken of this offer. By April 1941 a scheme of medical organisation had been drawn up and each Home Guard battalion was required to appoint a battalion medical officer with the rank of major; stretcher bearer squads were to be trained and a regimental aid post established at battalion headquarters. In addition, casualty evacuation schemes had to be prepared and an authorised scale of medical equipment fixed down to platoon level. A first aid exercise involving C Company of the 1st Stirlingshire Battalion on Sunday 14 September 1941 demonstrates that these arrangements found their way down to local units. Major J. Wright Wilson, the battalion medical officer, issued the orders and required platoons to produce designated numbers of stretcher cases, sitting and walking wounded.[57]

These initial arrangements were later modified by the appointment of zone medical advisers – a lieutenant-colonel's appointment – and by the appointment, where needed, of additional medical officers in the rank of captain or lieutenant, recognising that a battalion could be spread over a whole county and that medical and first aid provision had to be local to be effective. The 3rd Renfrewshire, for example, operating to the south of Glasgow in the Giffnock, Eastwood and Clarkston area had four medical officers on their strength.

Most Scottish battalions seem to have persuaded local general practitioners to take on the role of medical officer although few were graced with quite such a distinguished MO as the 2nd Aberdeenshire Battalion, whose officer roll included Professor Sir John Boyd Orr, FRS, an expert on nutrition and physiology, who, after the war, became the first Director General of the United Nations Food and Agriculture Organisation from 1945–48. Sir John, who was created Baron Boyd Orr in 1949 and received the Nobel Peace Prize the same year, had served in the Royal Army Medical Corps in the First World War, winning the Military Cross and the Distinguished Service Order.

In the university cities where there were medical schools, there was a plentiful supply of doctors and medical students and special arrangements were made to make best use of them through University Home Guard battalions. The 9th City of Aberdeen (University) Battalion, originally formed as a company-sized unit within the 4th Battalion, achieved battalion status in February 1943 and was organised in three rifle companies and M Company drawn from staff and students of the university's medical faculty. On stand-to, that is, on receiving word of a raid or invasion, the senior students would come under the command of the assistant director of medical services, North Highland District, and would be posted to reception stations throughout the Highlands to assist medical officers in dealing with casualties.

In May 1943 the Home Guard celebrated its third anniversary with a nationwide programme of events. At Buckingham Palace the Home Guard relieved the Scots Guards on public duties and were inspected by the king, who had taken on the appointment of Colonel-in-Chief of the Home Guard. The prime minister, in the United States for a conference with President Roosevelt, broadcast to Britain, praising the work of the Home Guard. Across Scotland parades and services and demonstrations marked the event. On Sunday 16 May the General Officer Commanding Scotland, General Sir Andrew Thorne, marched at the head of the Edinburgh Home Guard along Princes Street past the saluting base where the Lord Provost took the salute. General Thorne's action in marching with the Home Guard was widely welcomed as a tribute to the force from the regular army. In Alloa the Clackmannanshire Battalion showed that the "wee county" was up to speed in Home Guard matters; after a drumhead service, which was the first time the five companies in the battalion had paraded together, there were demonstrations of battle training and the use of sub-artillery. The day ended with the explosion of four anti-tank mines and the discharge of four barrel flamethrowers. As the *Alloa Advertiser* commented: "The resultant billowing cloud of flame and smoke mushrooming into the sky must have satisfied even the most fervent sensation seeker."[58]

At the end of 1943, with the Home Guard numbering around two million men and increasingly well equipped and well trained, it had to cope with damaging suggestions that the invasion threat was unreal and that so much training was a waste of effort. The prime minister took the opportunity of a question in parliament to emphasise the vital work that the Home Guard was doing in home defence and to repeat his view that "this was no time to relax any of our precautions or discourage our auxiliary services."[59] He did, however, say that he had asked that steps should be taken, compatible with public safety, to ease the strain which was being felt by efficient members of the Home Guard who were working long hours in civilian occupations.

Chapter 8

THE STRANGE CASE OF HAUPTMANN ALFRED HORN

It was not unusual for Home Guard units to be involved in the search for German airmen who had baled out over Scotland, to investigate reports of landings by boat or submarine, or to search for escaped prisoners of war. The Home Guard, with its network of units and sub-units in every part of the land, was well placed to put men on the ground where and when needed and at short notice. Many of these call-outs were false alarms, some indeed deliberately engineered by the Germans to cause confusion and panic. Back in the autumn of 1940, for example, German aircraft flying over Ayrshire had dropped empty parachutes and weighted satchels containing maps and photographs. Units covering remote areas were also often involved in the search for crashed aircraft, both allied and enemy.

Few of these incidents attracted quite as much notice as the crash of a German Messerschmitt Bf 110 twin-engined fighter and the parachute landing of its pilot who carried papers in the name of Hauptmann Alfred Horn of Munich. Hauptmann Horn landed near Eaglesham in the area of the 3rd Renfrewshire Battalion of the Home Guard. It was May 1941, early in the potential "invasion season" of that year. Britain and her Commonwealth allies had been at war alone against Germany and Italy since the fall of France a year earlier. A German invasion of Britain still seemed a real possibility, as did a German thrust into the Middle East with potentially fatal effects on British communications to the Far East and oil supplies.

On the night of 10 May the police in Giffnock, on the southern outskirts of Glasgow, advised the headquarters of 3rd Renfrewshire Battalion in the Scout Hall, Arthurslie Drive, Giffnock, that a plane had crashed near Eaglesham. Word was sent by the battalion's private telephone system to C Company headquarters at Busby who in turn notified an outpost near the spot. Two Home Guard officers, Lieutenants Cameron and Gibson, were on their way to a routine inspection of the outpost and had seen the parachutist leave the plane. On reaching the outpost they were briefed by the NCO in charge on the information that had been received from C Company. The two officers, together with two other ranks from the outpost then made for the scene of the crash, where they met Lieutenant Clarke from the Clarkston Company who lived at Waterfoot. Lieutenant Clarke had gathered some gunners from a Royal Artillery camp at Bonnyton Farm. On arrival at the scene of the crash the Home Guard officers were advised that a German airman was in a nearby farm cottage. Lieutenant Clarke then took the airman by car to C Company HQ, leaving the gunners mounting guard on the aircraft while the other Home Guardsmen searched the area for other parachutists who were presumed to have landed; the normal crew for a Messerschmitt Bf 110 was two, with three carried when on night fighter duties. No other crew member was found and it later emerged that Hauptmann Horn had flown solo from Augsburg, near Munich, and that the aircraft had been specially modified for solo use and fitted with additional fuel tanks.

By 23:45 3rd Battalion HQ had been advised that the prisoner had reached C Company HQ. While at C Company it was noted by Major George S. Helme, the company commander, that the prisoner was perfectly docile and observed strict military etiquette. He did however remark to Major Helme: "Are your Home Guard armed now?"[1] By 00:14 on the 11th Major Helme and the prisoner had reached battalion HQ.

Battalion headquarters had contacted the nearest army unit, 14th Argyll and Sutherland Highlanders, who rather washed their hands of the matter by advising that the prisoner should be lodged in Giffnock police station. The Battalion Commander, Lieutenant-Colonel Hardie, suggested to the 14th Argylls that the prisoner seemed to be of some importance, had been slightly injured on landing and should at least be in a military barracks if not a military hospital. He also reported that the prisoner had a message for the Duke of Hamilton and should be interrogated without delay as he seemed very willing to talk. The Argylls referred the matter up to sub area headquarters and Hardie was told that an escort would be sent from another unit in the area, the 11th Cameronians, to take charge of the prisoner and to replace the guard on the crashed plane.

Lieutenant-Colonel Hardie had formed the opinion that Hauptmann Horn was no ordinary pilot, noting that his uniform was new and of particularly good quality and had evidently not seen service. An observer corps officer arrived and identified Hauptmann Horn as either Rudolf Hess, the Deputy-Führer of Germany and Hitler's right-hand man, or his double. Hardie asked the prisoner if he was in fact Hess but the prisoner insisted that he was Hauptmann Horn. In the event Lieutenant-Colonel Hardie felt that it would be appropriate to treat the prisoner with an extra degree of courtesy and assigned a senior officer, Major James Barrie, commanding W Company, to accompany the prisoner to Maryhill Barracks. Eventually at 02:12 Lieutenant Whitby and a detail from the 11th Cameronians arrived at Giffnock to take charge of the prisoner; quite why it took so long to get Lieutenant

Whitby the two and a half miles from his base to the 3rd Battalion HQ is unclear. Major Barrie and Lieutenant Whitby, two armed escorts and the Deputy-Führer left for Maryhill Barracks, while the Cameronian detail went to guard the plane.

Maryhill Barracks had been warned to expect a German prisoner but when Barrie and his party arrived no preparations had been made, and the duty officer, 2nd Lieutenant Fulton of the Highland Light Infantry, was still in bed and, to quote Major Barrie's report he: "attempted to make all arrangements from his bed until advised that he should get up and dress. At no time did he give me the courtesy of my rank of Major, either by addressing me as 'Sir' or saluting me after he was fully dressed or when Lieutenant Whitby and myself finally left him."

Major Barrie was clearly unimpressed with what he saw of the efficiency of the regular forces: "It should be possible to get things done more expeditiously by having officers on duty dressed at their posts and also men standing by to act immediately. If the Home Guard can do this after doing their own work during the day it is surely more necessary that the regular forces should do so, otherwise with the speed of modern warfare many unfortunate incidents could be pictured occurring."

Lieutenant Whitby reported that the only accommodation offered was in the barracks' guard room. Hess protested at this and Major Barrie suggested that as the prisoner had some injuries he might more appropriately be accommodated in the barracks' hospital. Whitby also commented that Lieutenant Fulton never paid due respects to Major Barrie and seemed to lack "sufficient experience to cope with the situation". Admittedly a Deputy-Führer was hardly an everyday visitor to Maryhill Barracks but one could argue that the Home Guard had handled their end of the business with slightly more finesse than the regulars. A comment in the file from Colonel Campbell, Home Guard commander of Clyde Sub

Area, refutes to some extent Major Barrie's complaints about the inefficiency of Maryhill and Lieutenant Fulton by noting that the major was "inclined to make a mountain out of a molehill".

When the prisoner was examined by the Maryhill Barracks medical officer he was found to have hurt his right ankle, was experiencing some pain in the upper lumbar region and was suffering from gastric trouble of old standing. First aid was provided for the ankle and ipecacuanha, an emetic, prescribed for the gastric distress. At 14:00 hours on 11 May Rudolf Hess was removed from Maryhill Barracks and confined in Buchanan Castle Military Hospital, near Drymen. The local units of the 1st Dunbartonshire Home Guard stood to and maintained patrols for the two nights Hess was in custody here.

The chain of events that brought Rudolf Hess to Eaglesham has been the subject of a considerable literature but is hardly relevant to the story of the Home Guard. Hess was an experienced pilot who had flown in the First World War and regularly piloted himself around Germany. His message for the Duke of Hamilton, who was himself a pilot and an officer in the Royal Auxiliary Air Force, and incidentally the first man to fly over Mount Everest, was presumably some form of exploration of a peace settlement. As Hess flew to Scotland Germany was preparing to launch an attack on its ostensible ally the Soviet Union; Operation Barbarossa commenced just six weeks after Hess's flight. Hess had seen the Marquess of Douglass and Clydesdale (his courtesy title before he succeeded as 14th Duke on the death of his father in March 1940) at the Berlin Olympics when Clydesdale, then Conservative MP for East Renfrewshire, had visited Germany as part of a parliamentary delegation to observe the games.

The official explanation for Hess's planned visit to the Duke was given in the House of Commons by the Secretary of State for Air, Sir Archibald Sinclair:

When Deputy-Führer Hess came down with his aeroplane in Scotland on the 10th of May, he gave a false name and asked to see the Duke of Hamilton. The Duke, being apprised by the authorities, visited the German prisoner in hospital. Hess then revealed for the first time his true identity, saying that he had seen the Duke when he was at the Olympic games at Berlin in 1936. The Duke did not recognise the Deputy-Führer. He had however, visited Germany for the Olympic games in 1936, and during that time had attended more than one large public function, at which German ministers were present. It is, therefore, quite possible that the Deputy-Führer may have seen him on one such occasion. As soon as the interview was over, Wing Commander the Duke of Hamilton flew to England and gave a full report of what had passed to the Prime Minister, who sent for him. Contrary to reports which have appeared in some newspapers, the Duke has never been in correspondence with the Deputy-Führer. None of the Duke's three brothers, who are, like himself, serving in the Royal Air Force, has either met Hess or has had correspondence with him. It will be seen that the conduct of the Duke of Hamilton has been in every respect honourable and proper.[2]

When Sir Archibald was asked to comment on press reports that Hess had written a letter to the Duke before coming to this country, the secretary of state said: "I cannot speak to whether or not Hess wrote a letter to the Duke of Hamilton; I can only say that no letter from Hess to the Duke of Hamilton has reached the Duke or any responsible authority in this country." He declined to speculate on Hess's motives for seeking out the Duke.

Mysteries of course continue to surround the incident. Why

did Hess think that the Duke of Hamilton was a figure of such significance that it was worth risking a dangerous flight, which was intended to land at the Duke's private airstrip at Dungavel House, but which, due to the strip being unlit eventually required him to bale out over Renfrewshire? Was Hamilton merely an intermediary in German eyes and was Hess expecting eventually to meet a more politically significant figure? Why was his Messerschmitt Bf 110 not intercepted on its flight? Was the flight authorised by Hitler? Was Hess's flight part of a British intelligence plot to persuade the Germans that there was a peace party in Britain prepared to reach an accommodation with Hitler, the hope being that Hitler, feeling free from a Western threat, would attack Russia and bring the Soviet Union into the war? If, as seems possible, British intelligence was expecting a German envoy to fly to Dungavel, were they expecting Hess or somebody much less senior in the Nazi regime? Was the man captured by the Renfrewshire Home Guard really Rudolf Hess? Was the man who landed in Eaglesham the man who died in Spandau jail in 1987?

Chapter 9

ANTI-AIRCRAFT

One significant, but still surprisingly little-known, area of Home Guard activity was their role in manning anti-aircraft rocket batteries and in other forms of anti-aircraft defence. The lack of information about this is surprising as there were nine rocket batteries – or Z batteries, as they were originally known – manned by the Home Guard in Scotland, with an authorised strength of over 12,000 men. However the very existence of a British anti-aircraft missile system was not officially announced to the public until March 1944. Details of the Home Guard's work in rocket defence was not disclosed until September of that year and this official secrecy probably explains much of the lack of public awareness about the Home Guard rocket force.

When recruitment started in Scotland for Home Guard manned rocket batteries in April 1942 they were referred to as "a new and revolutionary type of anti-aircraft equipment".[1] It was noted that the equipment was still on the secret list and details of the weapon could not be given to men before enlistment and swearing-in. Scottish newspaper editors were given a demonstration of the Z rocket at an Edinburgh site in September 1943 but secrecy still prevailed. In November 1943 the *Scotsman* ran a story about Home Guard anti-aircraft gunners in the Edinburgh area and wrote that the "vague shapes of the guns, the details of which are not yet released to the public, could be seen against the sky."[2] The guns in question were twin-barrelled Z projectors.

Of the nine Scottish batteries, one was in Aberdeen, at the Links Golf Course; one in Dundee at Mid-Craigie; two were in Edinburgh, at Craigentinny Golf Course and at Crewe Road; two in Glasgow, at Blackhill Pits, Lambhill, in the north of the city and Prospecthill Road, Mount Florida, in the south; one at Hardgate Golf Course protecting the Clydebank shipyards; one at Ralston Golf Course near Paisley, overlooking the vital Rolls Royce aero-engine plant at Hillington; and the last at Greenock Golf Course defending the Greenock shipyards and harbour and the vital convoy anchorages at the Tail of the Bank.

The Home Guard batteries were formed in the spring of 1942, and along with the rest of the Home Guard were stood down at the end of 1944. However, the story of Britain's anti-aircraft rockets goes back to 1936 when a team of government scientists at Woolwich Arsenal began to look into rocket weapons.

In the years before the outbreak of the Second World War there was great concern about the threat posed by air attacks. The destruction of the Spanish city of Guernica by the German air force in April 1937 fuelled these fears and the view, expressed by Stanley Baldwin as long ago as 1932, that "the bomber would always get through" was widely held. Britain prepared for the coming war in the air by building modern fighters such as the Spitfire and the Hurricane and by developing a radar-warning network. Rockets were seen as a valuable form of defence against both low-level dive-bombers and high-flying night raiders. Trials were held in 1937 and 1938 at various sites in Britain, including the Western Isles. At the end of 1938 further trials were held in the more certain climate of Jamaica.

By August 1940 the first rocket batteries were being formed.

Regular soldiers of Anti-Aircraft Command were being trained in the operation of the new weapon, and engineering firms throughout the country received orders to produce the rocket launchers, with one large contract going to the Glasgow firm of Sir William Arrol & Co. Ltd. The rockets – officially known as UPs or "unrotated projectiles" (as opposed to a gun shell which was rotated by the barrel rifling as it was fired) – were fuelled by a 12½lb cordite charge produced at the Royal Ordnance factory in Bishopton, Renfrewshire. The rocket, measuring 6ft 4ins long and weighing 56lb could reach a height of 18,500 feet, travelled at a maximum speed of 1,500 feet per second and carried a 4¼lb high explosive warhead which was detonated either by an air pressure fuse, or later by a proximity fuse. The initial production of rocket projectors was of single- and twin-barrelled models, but in 1941 a nine-barrelled model was developed although the twin-barrelled projector was the most frequently found type.

Home Guard involvement with these weapons conveniently addressed two concerns. The first was the recognition of the changing role of the Home Guard as the invasion threat diminished. The second was the serious manpower shortage in the army. Anti-Aircraft Command – which manned heavy and light ant-aircraft guns, rocket batteries and searchlight batteries across the UK – employed huge numbers of men. In May 1941 the Command had just over 300,000 men on its strength. In order to release troops for the field force in the UK and in foreign theatres of operations, Anti-Aircraft Command was, in September 1941, ordered to reduce its strength by 50,000 men.[3] In a memo of 11 October General Pile, Officer Commanding Anti-Aircraft Command, wrote to the Officer Commanding Home Forces, General Alan Brooke: "I believe that night raiding is already on the wane and that the defence against it is becoming increasingly so effective as to render it eventually too dangerous a pursuit. When this day arrives I think the whole of ADGB [Air Defence Great Britain] should be turned into a part-time army (Home Guard) and that all my present show should be turned into a real AA umbrella for the Field Army."[4]

The loss of personnel was partly met by using women from the Auxiliary Transport Service – the ATS – in non-combat roles such as searchlight units; eventually around 74,000 women were engaged in anti-aircraft work. In October 1941 the idea was approved of manning rocket batteries with Home Guard personnel. The first such battery was formed in Liverpool and became operational in January 1942. These measures certainly worked; by 1945 Anti-Aircraft Command had as many heavy anti-aircraft guns in action in the UK as it had deployed in December 1940 but with 71,000 fewer regular gunners. Regular shake-outs of anti-aircraft gunners from Air Defence Great Britain to the field army proved possible as the Home Guard took up the task of manning anti-aircraft rockets and guns. By July 1942 it had been decided to apply compulsory Home Guard enrolment to man anti-aircraft units.

The Deputy Chief of the General Staff, Lieutenant-General Weeks, pointed out the great advantage of the rocket battery in this context: "The simplicity of the weapon lends itself to Home Guard or native manning" and "projectors are cheap and easy to maintain and can be manned by comparatively unskilled personnel."[5] He also noted in the same memo that the rockets were most valuable in obtaining a high density of fire on a definite lane used by enemy aircraft "such as is established when a fire is started in an industrial area and relays of aircraft come in to bomb".

Such conditions had been experienced in Scotland during, for example, the Clydebank Blitz of March 1941 and the Greenock Blitz of May 1941 and it was not long before Z batteries were being formed in the industrial areas of Scotland. The first battery to be formed in Glasgow, 101st Glasgow Home Guard Z Battery, was linked to a regular Royal Artillery rocket battery based at Prospecthill Road and recruitment for the battery was launched at a football match at the nearby Hampden Stadium. Edinburgh's first Home Guard Battery was formed from within the structure of the 4th Battalion of the Edinburgh Home Guard. One of the problems of asking for volunteers to sign up for anti-aircraft duties was that many of the likely volunteers would be young men who would soon be caught up by conscription to the armed forces.

Because the Home Guardsmen were volunteers, normally with full-time jobs to hold down during the day, the number needed to man a Home Guard battery, compared to a Royal Artillery rocket battery, was much higher. As a result these Home Guard batteries, each commanded by a major, might have as many as 50 officers and 1,400 other ranks in addition to a small cadre of regular gunners.

A typical Home Guard rocket unit usually worked a system involving a team of officers and men under a captain who were on duty one night in eight. The duty nights could therefore be rotated so that the same men were not always on duty on, say, a Saturday night. This, in theory, enabled the 64 twin-barrelled projectors in a battery to be manned. In reality, due to sickness and absenteeism something like 40 or 50 projectors were typically manned. One battery in Scotland, the 102 Glasgow, based in the north of the city, re-equipped and became operational on the less-common 9-barrelled projector in July 1943 and should in theory have been able to man 12 of these units; in practice a more usual turn-out was 10 projectors manned.

The eight-relief system allowed for the rockets to be manned every night in normal circumstances; however, provision had to be made for a situation when the Home Guard was mobilised in the event, for example, of an invasion. This would have meant the Home Guard becoming full-time soldiers and the Z batteries would have been grossly over-manned with eight reliefs available. In such, fortunately only theoretical, circumstances, the arrangement was that three-eighths of the battery strength would be retained on anti-aircraft work, allowing three-shift 24-hour manning, while five-eighths would revert to the infantry role with their parent link battalion. It was thus necessary that all anti-aircraft gunners should have at least the basic skills of the infantryman, both in the event of having to return to the parent battalion and to be able to defend the battery against ground attack.

In a typical battery such as that at Dundee, men reported for their turn on duty at 7.30 p.m., checked the equipment and carried out drills and training until 9.30 when an evening meal was served. They were free to rest and sleep until breakfast and the end of their duty spell at 6.30 a.m. Of course if there was a raid or an alert they had to stand-to and man their weapons and the chance of a quiet night's sleep would be lost.

Even this limited commitment of one night in eight caused problems for employers who were concerned that overnight Home Guard duties would leave men less able to perform a full day's work in industry. When the Dunbartonshire Z Battery was established at Clydebank & District Golf Club's course at Hardgate in April 1942 Sir Stephen Piggott, the Managing Director of John Brown's shipyard at Clydebank, wrote requesting that his shipyard workers should be exempt from any requirement to perform night-time Home Guard duties at the battery.[6]

The eight-relief system, as opposed to a seven-relief system, had advantages in moving duty nights but there were problems with integrating this Home Guard duty rota with work patterns. For example a man might have a requirement to do overtime on two fixed nights of the week and this would eventually clash with his Home Guard duties. Teams had to be trained together to reach maximum efficiency and men were trained in specific duties in the anti-aircraft role. It was impractical to train everyone to perform every role and if men turned up on a night when their regular relief was not on duty they had to be fitted into an existing team. There was in Scottish Command a factory liaison officer whose duty was to negotiate with industry over just this sort of problem. A regular army officer, serving as admin officer to a Home Guard battery "somewhere in Scotland" was quoted in the *Glasgow Herald* about this liaison work: "Meetings with factory managements arranged meetings with the foremen. The managements allocated clerical staff to handle the paper work which the battery staff could never have coped with. It meant a lot of trouble for the foremen and plenty of give and take, but method and good will is solving the jigsaw puzzle of factory and battery shifts."[7]

Despite compulsion, recruitment proved to be a problem for many of the batteries and in March 1943 a scheme was evolved whereby each rocket battery was linked to a Home Guard battalion and drew its personnel from it; thus the 102nd Glasgow Battery was linked to the 2nd City of Glasgow Battalion of the Home Guard and the 101st Glasgow Battery to the 3rd Battalion. An absolute priority was given to the staffing of Z batteries as opposed to the infantry units of the Home Guard. The transfers of first-rate men from general service units to rocket batteries was of course not always an easy matter and undoubtedly had serious consequences for the battalion which saw itself drained of its best men. The history of the 2nd City of Glasgow Battalion noted: "The Battalion was gradually bled white of its younger and more active personnel. Its effectiveness as an infantry general service unit was becoming seriously impaired, and the strain of duty on top of long hours of civilian war work was beginning to tell, especially on those original members now qualified to wear the 4-bar chevron denoting 4 years' service."[8]

The link battalion scheme was a great help but it was not a complete answer to the needs of the anti-aircraft Home Guard. A report of a visit in April 1944 to the Aberdeen Z Battery and Heavy Anti-Aircraft Battery revealed that the link battalion, 4th Aberdeenshire, had put up 94 men from A, B and C Companies of whom only 27 had been found suitable for anti-aircraft work, while out of 160 men from D Company only 24 were accepted for anti-aircraft duties.[9] The author of the report, Colonel Bowhill of 20 Home Guard Anti-Aircraft Regiment, noted that he had instructed his battery commanders to go to the limits of compromise as regards suitability for anti-aircraft work. There was, however, evidence of goodwill and willingness to make the system work. In May 1944 Major Reid, commanding the 71st Aberdeen Heavy Anti-Aircraft Battery, wrote to the Officer Commanding 4th Aberdeenshire Battalion: "The last intake that came over for Anti-Aircraft training were so well trained in General Service work that it was a pleasure for my officers to take them in hand for gunnery. Incidentally they passed out as gunners in record time and are as good as any in the battery."[10] And although the general principle of one link battalion in an area doing basic training and feeding men into anti-aircraft work was adhered to, this could be modified and in the Aberdeen case it was agreed to look at the possibility of drawing off men from the 7th Aberdeenshire (Works) Battalion, even though this unit had a variety of other demands on its resources.

As the war went on the purely voluntary nature of the Home Guard was modified and men were directed into anti-aircraft units by the Labour Exchanges. Many units felt that some of these later recruits were less enthusiastic and reliable than the original volunteers; although the 101st Glasgow Battery noted these problems the unit history also acknowledged that some of the directed men performed well and became non-commissioned officers.

Many of the batteries were to find problems with absenteeism; the 102nd Renfrewshire Battery at Greenock attributed this to the long and uncertain hours worked by the men in the local shipyards, and one can easily imagine that after a hard day's work in the yard the prospects of turning out on a winter's night to man an anti-aircraft battery on top of a hill would be less than appealing.

Morale could also not have been helped by the fact that by the time the Home Guard batteries were operational the main air attacks on Scotland were over. Most batteries only saw action on one occasion and some were never called to action stations. The 101st Edinburgh Battery at Craigentinny Golf Course fired on a Junkers 88 bomber on the night of 24/25 March 1943 and was credited with a shared kill, while the 102nd Edinburgh Battery at Crewe Road South was in action the same night and shared the credit for another kill with a Royal Artillery heavy anti-aircraft battery and a Polish anti-aircraft battery.

In Glasgow the Home Guard unit at Prospecthill Road fired 60 missiles at German raiders on the same night and shared the credit for a kill with other Z batteries and heavy anti-aircraft batteries in the area. However, the 102nd Glasgow Battery at Blackhill Pits was never called to fire its rockets in anger, an experience it shared with the 102nd Renfrewshire Battery at Greenock. The Aberdeen Battery went to action stations during an air raid on Aberdeen in April 1943 but could not fire its rockets as the German aircraft were flying too low and had, as a result, the deeply distressing experience of watching their homes being bombed without being able to respond or leave their post to render assistance. The 101st Renfrewshire Battery at Ralston Golf Course was called to action stations on 20 March 1943 but the enemy raider did not come into a suitable position for rocket firing.

Training presented a problem for these units; drills and dummy firing could be carried out at the battery sites but live practice firing was hardly advisable over a built-up area, even if most batteries were located on golf courses with open ground for rocket debris to fall on after the warhead exploded. Seaside locations were ideal for live firing, the Greenock unit travelling down the coast to Ardrossan and the Dundee unit firing from Easthaven, near Carnoustie.

In April 1944 Scotland's Home Guard Anti-Aircraft units were reorganised into two AA Regiments, the 20th with HQ in Edinburgh under Colonel Alexander Bowhill controlling the Edinburgh, Aberdeen and Dundee batteries, and the 21st with HQ in Glasgow under Lieutenant-Colonel Elphinstone Gillespie and having under command the two Glasgow, two Renfrewshire and the Dunbartonshire batteries.

The existence of the Z rockets might be a military secret but the batteries themselves could hardly be concealed from public view. The spectacular sight of a multiple launch would undoubtedly have impressed the civilian population as much as it intimidated enemy pilots. The arrival of a rocket battery of course had its impact on the local community; Clydebank Golf Course lost a number of trees, felled to clear lines of fire for the rockets of the 101st Dunbartonshire Battery, and the golfers now needed a security pass to get to their clubhouse.

Like all Home Guard units the rocket batteries enjoyed the services both of men with vast military service and of absolute novices. The Dunbartonshire Battery's first commanding officer was Major G. de C Findlay, VC, MC, a former regular officer of the Royal Engineers and amongst its officers it counted experienced artillery officers from the First World War. This battery had a policy that all officers, whatever their previous military service, should enrol in the ranks and qualify on the Z rocket before taking up their officer's appointments. The battery considered that nine weeks' training of four hours per week was needed to make the average man proficient at Z rocket drill.

The Greenock Battery noted that men working on Z rocket batteries needed a high standard of physical ability: good eyesight to see fuse scales, good hearing to hear the aiming and firing orders and strength to carry and load the 56lb rounds. The Clydebank Battery noted that many shipyard workers, especially caulkers and riveters, had lost so much of their hearing due to the noise of the yard machinery that they could not hear commands and were unsuited to this particular form of Home Guard service.

The part-time volunteers of Scotland's Home Guard rocket batteries carried out a faithful watch on the skies for over two years, manning over 400 rocket projectors from Aberdeen to Greenock. As the Edinburgh *Evening Dispatch* commented, when giving its readers the first details of the Home Guard's involvement in rocket batteries, the 101st Edinburgh Battery had completed 820 nights of continuous manning.[11]

Scotland's Z batteries may not have destroyed many enemy planes, although their success rate when called upon to go into action was remarkably high. The nine batteries totalled three live firings and were credited with three shared kills.

Nor, as we have seen, were rocket batteries the only form of anti-aircraft defence in which the Home Guard took part. Heavy anti-aircraft guns were manned by Scottish Home Guard units. A battery was established at Aberdeen and was fed men by the 4th Aberdeeenshire Battalion. The Lord Provost of Aberdeen fired the first round from the Home Guard's 3.7-inch gun and commented: "It's lucky ma lugs were pluggit!"[12] By October 1943 this battery had enough trained men to take over eight 3.7-inch guns that were located at two sites in the city.

At Dundee an independent troop of heavy anti-aircraft artillery was formed in January 1943 from two platoons of the 1st City of Dundee Battalion and was incorporated into a Royal Artillery HAA battery. When the Home Guard gunners were passed as proficient they were given the manning of two guns.

In the Clyde area the 71st Clyde HAA Battery was established in December 1942 with its constituent troops in Lanarkshire, Renfrewshire, Dunbartonshire and Ayrshire. In March 1943 a Glasgow Troop was created; drawn from the works unit of G & J Weir's engineering complex, these men had previously been part of a works unit in the 5th City of Glasgow Battalion. The Lanark Troop, based at Baillieston, engaged an enemy aircraft in the spring of 1943. In 1944 it was decided to add a second two-gun section to each troop; previously only the Ayr Troop had two sections, and a second troop was added in Dumbarton. With these additions it was determined to split the battery into two and the 72nd Battery was formed. Men were drafted into these new sections and troops from general service battalions and training commenced. However, shortly afterwards Home Guard parades were put onto a voluntary footing and the Home Guard was stood down.

In addition to rockets and heavy anti-aircraft guns the Home Guard also manned light anti-aircraft units across Scotland. These were equipped with a mixture of 20mm cannon and various machine-guns. These light anti-aircraft troops were manned by Home Guard general service battalions and were chiefly located at significant defence production centres. For example the 1st Dunbartonshire Battalion controlled light anti-aircraft troops at the Royal Navy torpedo factory in Alexandria and the Blackburn aircraft factory in Dumbarton. The 2nd Scottish Borders Battalion had a light anti-aircraft troop at ICI's Charlesfield bomb factory near St Boswells. This factory employed over 1,000 staff and turned out a million incendiary bombs a month. Eight 20mm guns protected the installation manned by workers who were serving in the 2nd Scottish Borders Battalion and other neighbouring units.

The role of the Ministry of Aircraft Production in making anti-aircraft and other armaments available to aircraft and related industries was touched on in Chapter 3. A War Office file on the Home Guard AA Defence of Factories in West Scotland[13] shows a fairly confused pattern of provision with a varied mix of weapons being provided by a wide range of sources; for example, six .300 Lewis Guns at the Barr & Stroud optical factory in Glasgow came from the Ministry of Aircraft Production; three Bren guns at the Royal Ordnance factory, Bishopton, had been sourced by the Ministry of Supply, and two .303 Hotchkiss machine-guns at the Arrochar torpedo range were provided by the army ordnance depot at Stirling. Matters were somewhat regularised in February 1943 when Scottish Command laid down two schemes for Home Guard LAA troops. In all areas other than Edinburgh and West Scotland troops would be formed at vital points as independent units under command and administration of a nearby anti-aircraft unit. In Edinburgh and West Scotland the LAA troops would be sub-units of General Service battalions and would operate in daylight hours if and when there was a raid. Under this scheme the ICI explosives works at Ardeer would have twenty-four 20mm cannon and the Alexandria Torpedo Factory twelve light machine-guns. However, the plan to have independent units was set aside and the West of Scotland pattern became the UK norm following an army council order of November 1943. This decreed that Home Guard LAA troops would be formed by and within Home Guard battalions who would be responsible for their administration: "The primary role of a Home Guard LAA Troop will be the defence of their factory or establishment during the hours of daylight within the alert. As members of their parent battalion personnel of these LAA troops will have a secondary ground defence role in an emergency which renders impossible the exercise of their primary role."[14]

In 1944, when one might have thought the risks were diminished, many of these west of Scotland sites would have their defences increased. The Alexandria torpedo factory would trade in their light machine-guns for six 20mm cannon and the Kilmarnock ICI plant would give up its twenty-four light machine-guns for six 20mm cannon and two quadruple Marlin .300 machine-guns.

Whether operating rocket batteries, or heavy or light anti-aircraft guns, there is no doubt that the Home Guard made a significant contribution to the defensive umbrella over the United Kingdom; this work, although demanding, had its attractions for many Home Guardsmen. The role of general service battalions in the light of the decreased invasion threat and the absence of any German sabotage or disruption raids became increasingly unclear to many members, although strenuous efforts were made by everyone from the king and the prime minister down to emphasise the value and role of the Home Guard as a local defence force. However, the idea of defending one's home or workplace from the perhaps more real threat of a German aerial attack was an attractive one, and many men could identify with and commit to the task.

General Sir Frederick A. Pile was GOC in C Anti-Aircraft Command until April 1945. Writing his *London Gazette* report on the anti-aircraft defence of the United Kingdom he observed:

"Disciplinary control over members of the Home Guard was virtually impossible and it was an easy matter for those who were so inclined to evade all duty. It was due entirely to the service given by the unselfish that the Rocket Batteries became and remained a force which the German aircrews treated with the utmost respect."[15]

Chapter 10

A FLEXIBLE FORCE

Most Home Guardsmen in Scotland performed the normal infantry roles of drilling and training, and carried out regular patrols such as the beach patrol maintained by the 2nd Moray Battalion on the coast at Findhorn. Each night two patrols of two
men armed with Sten guns slept in a fishing bothy and at first
light carried out their search for signs of enemy agents being
landed or suspicious objects on the beach. However others, as
we have seen with the anti-aircraft Home Guard, branched out
into a variety of forms of duty. On Tayside in July 1940 the 4th
Perthshire formed a naval platoon to patrol the river and to act as
troop ferries. Between 10 and 15 men and 5 or 6 vessels ranging
from a salmon fishing boat and a dinghy with outboard motor
to a speed boat were recruited to this service. The Royal Navy
units stationed at Dundee did not operate above the Tay Rail
Bridge and so the long navigable stretch of the River Tay running alongside the Carse of Gowrie was patrolled by the men of
the Home Guard. In Edinburgh the 2nd Edinburgh maintained
a motor boat patrol on the Union Canal. On the Moray Firth
men of the 4th Inverness-shire Battalion, based in the burgh of
Inverness, operated a motor boat patrol as their contribution to
mine-watching in the waters of the Firth.

In parts of Scotland horseback patrols kept watch on remote

upland areas. The 4th Scottish Borders Battalion used a mix of horse patrols and hill observers: "These men, numbering about thirty, covered an area in the Cheviot Hills, stretching from Lustruther, west of the Carter Bar, to Kirk Yetholm, their boundary being the international boundary. Each man had his special route or 'Ride' as it was called, and carried out a 'Dusk and Dawn' patrol; these patrols were co-ordinated with the shepherds who were the hill observers. It is worthy of note that nothing has escaped notice in these hills since the formation of this Organisation, even regular troops training in the area have always been reported."[1]

The 4th Battalion had an establishment of seventeen horses and similar sized units operated in the 1st Scottish Borders Battalion area (Roxburghshire) and the 2nd Battalion area (Selkirkshire), all upland areas with a strong horse-riding culture expressed in local traditions such as Common Ridings. Naturally a unit could not simply decide to establish a mounted patrol. In May 1943 the 3rd Dumfriesshire through the GOC Scotland approached Commander in Chief Home Forces for authority to create a patrol of one NCO and twelve other ranks at Waterbeck for the hill country to the north and north-west. The costs involved would have been 15 shillings a month (£0.75) per horse for maintenance of horse and equipment where saddlery and harness were supplied by the owner, and 10 shillings a month (£0.50) where the War Office supplied the saddlery and harness. Home Forces had no funds to meet this sum and it would have necessitated a special approach to the Treasury for additional funds to meet the annual expenditure of £117. Home Forces asked Scottish Command not to pursue the matter, as they considered it unjustifiable in operational terms. Although the battalion covered an area of 300 square miles the official view was that "the risk of an airborne raid on that part of the country can be discounted."[2] Fortunately for the enthusiasm of the Dumfriesshire horsemen, by September Home Forces were able

to move matters on; as 156 horses had been struck from establishments elsewhere in the United Kingdom and only 77 new
horses had been authorised they were prepared to approach the
War Office with the 3rd Dumfriesshire's case. On 12 October
1943 the War Office duly authorised the Dumfries mounted
patrol.

At times it is hard to avoid the impression that the Home
Guard was used to take on tasks that nobody else wanted and
that were arguably of quite minor importance. In 1942 three
armoured trains, originally operated in Scotland by the British
army and then transferred to the Polish army, were found to be
surplus to Polish requirements. These trains, named "J", "K" and
"L", were dispersed around Scotland and the existence of Home
Guard units recruited from railway employees made a natural
home for them. The "J" train was originally destined for 8th City
of Glasgow (LNER) Battalion, while "K" would be manned by
the 10th City of Edinburgh (3rd LNER) Battalion, and would
be based at Saughton Siding, near Corstorphine but deployed to
Drem, East Lothian at action stations. "L" train went to the 4th
LNER Battalion, a unit in the Aberdeen area formed from LNER
personnel. For a variety of reasons the "J" train ended up in Fife,
stationed at Thornton Junction near Dunfermline, on the books
of the 5th Fife Battalion, and the "L" train was eventually manned
by the 7th Aberdeenshire Battalion, a works unit incorporating
railway personnel, and based at the Inverurie depot but deployed
to Kittybrewster in Aberdeen on action stations.[3]

In their Home Guard role the trains were each initially armed
with two Hotchkiss 6-pounder cannon, a Vickers machine-gun
and four Bren guns. They were manned by 16 Home Guardsmen under a captain, with a lieutenant as weapons training officer, two signallers and a train crew of four. In addition there was
a mobile base consisting of a passenger coach and brake van to

provide crew transport and catering. This came under the charge of the second in command, along with a company sergeant major, a train crew of three and a fighting crew of six – deploying two Bren guns and a Boys anti-tank rifle – an armourer and a further three men, making a total in the mobile base of 14. Thus each armoured train had a total complement of 38 officers and men. The mobile base would be detached from the train when going into action.

A description of the "L" train exists in a pamphlet written anonymously by a member of 4th LNER Battalion, a Home Guardsman who regretted that, acting on the old soldiers' motto of "never volunteer", he had not signed up to serve on it. He wrote: "This is a train of unbelievable formation being the unlikely combination of a camouflaged GE 2-4-2 tank and a large double bogie St Rollox tender with concomitant ammunition, HQ, and anti-aircraft wagons and armed fore and aft with a lethal weapon in the form of a rather outdated naval piece."[4]

The role of these trains was seen as threefold: to patrol coastal areas in order to locate enemy detachments and gain information, to reinforce a threatened point with firepower and to deal with attacks when the railway line offered the best or only means of approach to the line of advance of enemy tanks.

How effective an armoured train was likely to be in its patrol and information-gathering role was perhaps questionable and much would also depend on its ability to communicate any information it gathered to surrounding forces and higher command. A signals officer from Fife Sub Area inspected the "J" train in 1942 and found that the Home Guard's standard of Morse signalling was so poor that consideration was given to the use of pigeons to communicate with higher command. However it was decided instead to concentrate on improving the crew's signalling skills.[5]

There were evident problems in integrating the armoured

trains with the local Home Guard units; a company commander of the 2nd Fife Battalion wrote to his battalion headquarters pointing out that the "J" train had been in his area and suggesting that he be notified the next time it was there "so that I and my men may have a look at one of our Reserves with whom we might have to operate."[6] This request was granted by his CO and accepted as good practice by Fife Sub District who noted on the letter: "Remember for future and notify units when train passes through their areas."

Apart from routine static training and a patrol run of around 100 miles at least once a month along a route and at a time scheduled to avoid heavy traffic, the armoured trains were also used on special exercises. The history of the 10th City of Edinburgh Battalion notes that in September 1942 the "K" train, manned by No.2 Platoon, A Company, took part in a "large scale military exercise in Fort William/Spean Bridge area and the officers and other ranks were complimented on the very excellent manner in which they carried out the tasks allotted them."[7]

In February 1944 the *Scotsman* ran an article about the Edinburgh-based "K" train. The train, it revealed, was commanded by Captain W.J. MacLeod, whose day job was driving the "Flying Scotsman" train on the Waverley to Newcastle leg of its journey between Edinburgh and London. The crew, drawn from the 10th City of Edinburgh (3rd LNER) Battalion, included officers and men who in civilian life were drivers, firemen, signalmen, platelayers, electricians, engineers, traffic-control men, clerks, guards and a stationmaster. Platelayers and spare rails and sleepers were carried in case of the need for emergency repairs in action. The train had its engine in the middle, with a gun truck armoured with half-inch plate front and rear. The correspondent noted that if the fighting members of the crew had to leave the train to go into action they did so by trapdoors in the floors of the armoured trucks, thus presenting as few targets as possible to the enemy. His verdict was that armoured trains could be useful in mopping up diversionary enemy parachute attacks after the second front had opened and allied forces had invaded north-western Europe.[8]

Despite this view, perhaps the most realistic assessment of the armoured train appeared in the history of the 5th Fife Battalion which noted that: "every Sub District Commander who inspected it thought it a novelty without any real value and appeared to be relieved when told it was under him for administration purposes only."[9]

Just as the skills of railwaymen were utilised in the manning of the three armoured trains, so lorry and coach drivers across Scotland were incorporated into the Home Guard. Back in 1940, 600 passenger coaches and their drivers operated by the Scottish Motor Traction (SMT) group of companies were earmarked by the military authorities for use in an invasion emergency. The SMT bus companies controlled almost all passenger road transport outside the major cities and were well represented in priority areas. In 1941 Scottish Command raised Home Guard motor coach companies; 16 of these were formed with 1,850 men and made up an often very efficient part of general service Home Guard battalions. In September 1942 these companies were removed from their local Home Guard battalions and formed into the 1st Motor Coaches Battalion, under Lieutenant-Colonel James Amos, who in civilian life was director and traffic manager of SMT. The battalion took on its own infantry training, with specialised training being provided by local Royal Army Service Corps staff across Scotland. The existence of these motor coach units provided the opportunity to train Home Guard battalions in troop transportation and allowed battalions to extend the scope of their training by permitting them to move to different types of terrain and to use training facilities like sub-artillery firing ranges and grenade ranges that were distant from their bases.

A further reorganisation came in January 1943 when the Motor Coaches Battalion was formed into No. 1 Scottish Home Guard Transport Column. The column's headquarters were in Edinburgh and it had 12 companies widely distributed across Scotland, each with 60 coaches and a 10 per cent operational reserve to allow for unserviceable vehicles. At action stations the column could turn out fully equipped in four hours and had the capacity to lift 20,000 men. Part of this lift capacity was earmarked for general military purposes; for example 2169 Company based in Anstruther was committed to the transport of a Polish parachute brigade. The companies were armed with Sten guns, grenades, anti-tank rifles and Lewis machine-guns. Lieutenant-Colonel Amos received the OBE in the "stand-down" honours list.

The value of a personnel transport facility being apparent, it was decided in November 1942 to raise a second Home Guard transport column, this time consisting of load-carrying vehicles. These were lorries of 3- to 5-ton capacity organised in 12 companies around Scotland with the column headquarters being at 24 Queen Street, Glasgow, under Lieutenant-Colonel G.K. Crichton. By January 1944 No. 2 Scottish Home Guard Transport Column had 66 officers, 118 NCOs and 735 drivers, all of the latter being professional drivers in civilian life. The Shetland company of the column was transferred, for administrative reasons, to the 1st Zetland Battalion.

Most of the work of the Home Guard was very public and the reassurance that was given to people by the presence of a disciplined armed body of men in their local community should not be underestimated. While huge numbers of British and allied troops were stationed around the country there were areas where regular troops were thin on the ground and where the very public presence of the Home Guard provided both an assurance of military protection and an aid to civil power in matters such as fire-watching and rescue services.

However there was a little-known, covert and unacknowledged aspect to the Home Guard in the shape of the Auxiliary Units (AUs). These were secretive undercover units located in rural areas whose mission was to go underground in the event of an invasion and conduct a campaign of sabotage and irregular warfare behind enemy lines. Three battalions of these Auxiliary Units were formed with 210 (GHQ Reserve) Battalion, Home Guard, covering Scotland and Northern England.

By the nature of the Auxiliary Units' mission they operated in small cells or patrols of a handful of men; thus the Western Isles had seven patrols covering the area from Stornoway to South Uist, while in the 1st East Lothian battalion area one officer and twelve men served in Auxiliary Unit patrols. Each group was co-ordinated by a full-time intelligence officer, usually a regular infantry officer, who also liaised with the local Home Guard and army structures. Because of the secret nature of the AUs and the extremely dangerous nature of their post-invasion duties there was a tendency to recruit men who knew each other well or were relatives and who therefore had an existing bond of trust. In a patrol in Caithness there were, from the village of Keiss, a Sergeant Bain, a Corporal Bain and a Private Bain, and in Camster a Sergeant MacKay and two Privates MacKay. The same pattern of presumed kinship recruitment is found in many other localities.[10]

The recruitment of men for Auxiliary Units was not an easy matter and the covert nature of their role must have placed many of the AU Guardsmen in an embarrassing situation. They would have had to explain to friends and neighbours why they were no longer serving with other fit men in the area in the local battalion and to say where they went to so secretly and without apparent good reason – another excellent reason to recruit within a close family or circle of friends. In April 1941 the Home Guard Directorate had issued a "Most Secret" instruction about AU recruitment: "Only reliable men of discretion are enrolled, and I am therefore to request that every assistance may be given by HG Commanders to Officers of Auxiliary Units in securing the right men for this duty though it is realised that it may mean the loss of a good man to the local Home Guard unit."[11]

Requests for transfer of key men from General Service battalions to Auxiliary Units were often contentious. General Thorne, GOC Scotland, had issued an instruction that he placed great importance on the AUs in the defence of Scotland and that if officers were approached regarding the transfer of men to the AUs they should do all in their power to assist. The AUs naturally picked the best men for the job, who equally naturally were often key men in their existing battalions. In November 1943 an AU officer called on Lieutenant-Colonel Bain, commanding 2nd Banffshire Battalion, to seek the transfer of Lance-Corporal A.M. Wilson from B Company of 2nd Banffshire to 210 (GHQ Reserve) Battalion, the umbrella unit for Scottish AUs. Wilson's brother was already serving in the AU and Lance-Corporal Wilson was the only man in the area with the appropriate attributes and qualities. Bain interviewed Lance-Corporal Wilson, who was keen to transfer, but his company commander felt that he could not be spared. Bain therefore attempted to block the transfer on the grounds that his battalion was under strength and that when the problem was solved by compulsory direction he would have another NCO trained to replace Wilson and then transfer Wilson to the AU. However Bain's attempt was firmly crushed; a staff officer from Moray Sub District wrote on 30 November: "Whilst the Sub-District Commander realises your position in this matter he directs that you shall carry out this transfer forthwith in accordance with the instructions in the GOC's letter."[12]

Although these AU members were Home Guardsmen they were for all practical purposes separated from the normal Home Guard procedures and processes. For record purposes and allowances they were initially carried on the books of the Home Guard battalion HQs and information about the existence and location of Auxiliary Units was only allowed to be divulged down to the level of company commander. A War Office memo of 20 January 1941 had stressed: "It is important that the existence of these units should be mentioned as seldom as possible."[13]

Problems could arise from the secretive nature of the AUs. Lieutenant-Colonel Noble, commanding 1 Banffshire Battalion, wrote to Moray Sub District on the difficulties caused by the mutual non-recognition of Home Guards and AUs and suggested steps needed to be taken to make sure that they did not end up firing on each other, especially during hours of darkness. He mentioned the need for greater liaison and personal contact between the two forces: "At present, except perhaps in a nebulous way, their existence is unknown to many members of this Battalion as the information given to me about a year ago by their Commander as to their scope and purpose was not allowed to go below HG Coy Commanders."[14]

Arising from this letter a staff officer in Moray Sub District in May 1944 wrote to an officer organising AUs in that area authorising him to call on Home Guard battalion commanders to tell them of the role and location of AUs in their area but concluded: "The extent to which you can divulge the role of the Home Guard Aux. Units must be left entirely to you but it is appreciated that you will not be in a position to give more than a general indication and that you are strictly limited by Security considerations."

In all there were probably fewer than 1,000 men in Auxiliary

Units in Scotland, mostly in rural areas of the north, north-east, Borders and Dumfries where there was rough country and a chance to operate in a clandestine manner after an invasion. One essential part of their work was to prepare hideouts and stores where arms, ammunition and supplies could be cached and where the men could live and rest while engaged in their clandestine struggle with the enemy. When the Auxiliary Units stood down, along with the rest of the Home Guard, these hideouts had to be cleared, a task which seems not to have always been dealt with promptly. Captain W.A. Mustard, an AU officer, received an order from Aberdeen Sub District about a hideout in his area: "As place is locked it is presumed that it is stocked with ammunition. Maj MacNicol GSOII Aux Units to whom matter was referred has asked this HQ to get in touch with you in order for this place to be emptied and stores which may be there to be returned to appropriate authority. MacNicol agreed that there may be other hideouts in your area which have not been cleared although instructions had been issued for this to be done."[15] Captain Mustard was ordered to clear all the hideouts in the district, return the contents to the appropriate authority and certify to Aberdeen Sub District that this had been done.

In 1941 official attention began to turn to the possible savings in regular manpower that could be achieved by using Home Guard personnel in the many batteries of artillery that protected our coasts and harbours. There were initially two possible avenues felt worth exploration: one was to use the Home Guard as members of gun teams, the other was to use Home Guards as infantry defence platoons to protect coastal artillery sites. There were very real objections to the first due to the high state of readiness of these artillery batteries and doubtful availability of Home Guards to provide round the clock manning in normal circumstances, but the local defence role was seen as less contentious. In fact some artillery units had already been working with their local Home Guard battalions. The officer commanding 543 Coast Regiment, Royal Artillery, covering batteries from Dundee to Montrose, reported that when in 1940 at Broughty Castle, Dundee, he had started training members of C Company, 1st City of Dundee Battalion, there had been problems finding loading numbers for gun crews. However, layers and other specialists were soon proficient and all personnel were very keen. Unfortunately this initiative was stopped on the orders of the Home Guard zone commander.[16]

In September 1941 the commander of 542 Coast Regiment at Aberdeen reported that the three batteries at Torry Point and Girdleness, Aberdeen and Salthouse, Peterhead could each take on a 30-man Home Guard defence platoon. The 6-inch guns at Torry were kept at too high a state of readiness to permit the use of Home Guards but the others could use some Home Guard personnel while operating at their currently reduced manning levels. The commander said the Home Guard could also be deployed on the secondary armament of 4.5-inch howitzers and 75mm guns located at Torry.

A survey of Scotland's other coastal artillery units showed a mixed response to the idea of Home Guard manning, although in many cases this was due to a lack of a nearby Home Guard unit of useful size. The lieutenant-colonel commanding Clyde Defences who, apart from the Cloch Battery near Gourock commanded batteries on the north shore of the Clyde at Ardhallow and Toward near Dunoon, noted that there were no convenient local Home Guard units for these batteries but that his 12-pounder gun in Castle Gardens, Dunoon, which had been installed in 1940 to protect the anti-submarine boom, had a protective detachment of six Home Guardsmen. The Cloch Battery, mounting two six-inch guns, felt that six Home Guards could be used as gun crew but would welcome a 30-man defence platoon.

The Royal Artillery major commanding at Broughty Ferry took advantage of the changed attitudes and had the 1st City of Dundee Battalion doing night guard on his battery from 8 p.m. to 6 a.m. two nights a week. This allowed the Royal Artillery personnel rest and the local Home Guard commander was enthusiastic about increasing these guard duties to three turns a week.

By early 1942 things had moved on and the Cloch Battery was training 20 men from the 1st Renfrewshire Battalion as gun numbers and had the services of a further 50 for local defence. Colonel Armstrong, commanding Forth Fixed Defences, reported sixty Home Guard volunteers had started training at the Leith Docks battery and felt that they would form a valuable reserve, especially as they had agreed to give full-time service on "stand-to" or "action stations" and to go wherever they were most needed at such times. On the other shore of the Forth there were problems finding men to serve at Kincraig Battery located some two miles from the village of Elie. Men often felt that they had joined the Home Guard to defend their homes and were reluctant to serve several miles away at an isolated battery.

In a move analogous to that taken with Anti-Aircraft Command, steps were soon taken to reduce the regular complement of some batteries. Coastal batteries were placed in two categories: Land Service batteries which were the permanent pre-war batteries covering key harbours and installations, and Special Coast batteries, which used a pair of naval guns and searchlights and had been established during the war to cover less important ports or to enhance the protection offered by Land Service batteries. In 1942 GHQ Home Forces instructed both categories of batteries to take steps to have in place a Home Guard protective platoon of 30 men, with the expectation that 20 men would be available on stand-to. In addition Special Coast batteries were ordered to replace regular soldiers with about 20 Home Guards per battery,

but instructed that twice this number should be trained to provide adequate cover. By June 1942 Home Forces, anxious to save
manpower, ordered the closure or Home Guard manning of a
number of coastal batteries in Scotland and arrangements were
made for the compulsory drafting of Home Guardsmen to coastal defence batteries. These batteries, were, like the rocket batteries, given priority over general service units, and indeed where a
conflict of priorities arose between anti-aircraft and coastal artillery the manpower needs of the coastal batteries were ordered to
take precedence.

The Fort George battery would, it was noted in July, be manned
from 4th Inverness-shire Battalion, the Lossiemouth Battery
from 1st Moray Battalion, the Stannergate Battery from 1st and
2nd Dundee Battalions and the Pettycur Battery, near Kinghorn,
from 5th and 8th Fife Battalions. The Pettycur Battery retained
a regular artillery component and the 8th Fife element of Home
Guard manning was transferred to the 5th Fife Battalion. The
situation in Edinburgh District was not reported on. By August
1942 a return showed the situation to be as follows:

Battery Location	Home Guard Manpower
Stannergate, Dundee	20 HG trained, 18 part-trained, 35 recruits
Fort George, Inverness-shire	0 HG trained, 27 recruits and 53 still to be found by Inverness Sub Area
Lossiemouth, Moray	9 HG part-trained, 71 recruits commencing training 24 August
Dirleton, East Lothian	80 HG trained
Pettycur, Fife	21 HG trained, 17 part-trained and 30 recruits

By September matters had moved on quite satisfactorily and the return for that month showed the position as follows:

Battery Location	Home Guard Manpower
Stannergate	3 officers and 80 other ranks
Fort George	1 officer and 30 other ranks
Lossiemouth	2 officers and 58 other ranks
Direlton	3 officers and 105 other ranks
Pettycur	1 officer and 81 other ranks

By this time a number of other coastal artillery sites had been identified for Home Guard manning including Wick, Peterhead, Girdleness (Aberdeenshire), Montrose, Stranraer and Ardhallow (Argyllshire). It was reported on 28 September 1942 that the Girdleness Battery had no Home Guard personnel in place and that two officers and 80 other ranks were needed. These would be found by a variety of methods including voluntary transfer from other units, and voluntary or compulsory enlistment of new men, although it was hoped that ex-gunners from Home Guard units would take the opportunity for voluntary transfer. The Girdleness Battery was to be linked to the 4th City of Aberdeen Battalion; later on this linkage was transferred to the 7th City of Aberdeen Battalion.

In January 1943 Lieutenant-Colonel F.G. Holbrook, commanding officer of 501 Coast Regiment, Royal Artillery, reported on the problems of Home Guard manning. A typical site, he wrote, would have two officers and 29 other ranks from the regular forces and 80 Home Guard all ranks. These men would be required to man the following:

Weapon or equipment	Personnel needed
6-inch guns x 2	36 men

Searchlights x 2	4 men
Searchlight engines x 3	3 men
Field gun 1 or 2	8 or 16 men
AA rocket projectors x 2	4 men
Battery staff	5 men
AA and ground defence	As needed

He concluded that 80 men were needed to fight the battery and provide a protective screen for the site but that the full strength was not needed on action stations for a short engagement. If, say, 45 Home Guards turned up then that would be acceptable provided they were fully trained and the rest turned up and were also fully trained. It is very clear from Holbrook's report that these batteries could not be fought with just the Royal Artillery component of their establishment. There were questions about the level of turnout for training – a spot check at Fort George had shown that 24 men had been absent from training in December – although the keenness of the minority was noted and appreciated. On 27 January 1943 a trial "call-out" was held at the Innes Links Battery, Lossiemouth and 47 officers and men reported within 75 minutes, many of them cycling up to 7 miles on a moonless night over roads and tracks badly cut-up by construction work.

Holbrook would go on to report on unsatisfactory attendance at the Wick Battery and noted that the Caithness Battalion Commander was going to instigate prosecutions for absenteeism. He argued for a higher Home Guard manning level to allow for efficient manning with trained men. To this end he welcomed the transfer of the Ardersier Platoon of the 4th Inverness-shire Battalion to the Fort George Battery, giving it a Home Guard strength of 130 men, and planned to press for similar transfers in his other batteries in the north of Scotland. Holbrook's actions clearly had a beneficial effect, because when Lord Bridgeman, Director General of the Home Guard, visited Wick Battery in March 1943 there were two Home Guard officers and 51 other ranks on parade. Holbrook was able to report to the Commander Royal Artillery at Scottish Command that Bridgeman had: "expressed the view that, on the basis of what he saw, there was every possibility of the Home Guard attaining an efficient standard in a reasonably short time."

In October 1943 the Home Guard had evidently reached a fair degree of competence; a report in the *Scotsman* told of Home Guards having a successful night-firing practice at a 6-inch gun battery on an island in the Forth:

> The Colonel commanding the Coast Artillery Regiment in whose area this Home Guard battery is stationed, said to them after the firing was finished: "It is very satisfactory indeed, especially as it was, for so many of you, your first shoot. I am very glad that this was an entirely Home Guard 'show,' and you have shown that you have been well trained."

> In describing it as an entirely Home Guard show, the Colonel was referring to the fact that not only were the guns manned by Home Guard personnel, under the orders of their Battery Commander, but the searchlights were also in the hands of the Home Guard. At the beginning of the War the searchlights and the engines on these fortified islands were manned by the Royal Engineers, but with the demand for sappers in the Field, Coast Artillery gunners took over the management of searchlights and engines. Consequently the Home Guard, before beginning night firing, had to produce from their own ranks searchlight specialists and engine attendants – men whose civilian

jobs were connected with machines or electrical appliances naturally being the first choice for this job.[17]

The *Scotsman's* war correspondent went on to comment on the problems the Home Guard unit had faced in reaching and maintaining their efficiency. Changes in employment conflicting with training, younger men being called up to the armed forces, the movement of older men in reserved occupations to other parts of the country – all had presented difficulties with a recruits, class having always to be kept in being and an effort made to keep the interest of the trained men by broadening their skills, from artillery work to map reading, junior leader training and infantry weapons training. He also made the point that coastal artillery work demanded a high standard of skill and fitness; a six-inch shell weighed 100lb and had to be manhandled into the breech of the gun.

The Home Guard gunners undoubtedly made a significant contribution to coastal defences. The 5th Fife Battalion's history notes that their coastal artillery sub unit trained at Pettycur battery twice a week and carried out regular practice firings of both sub-calibre and full-bore charges in daylight and, under searchlights at night, using tug-towed targets. Such was their effectiveness in this role that the two 6-inch coast defence guns at Pettycur were, from December 1943, manned during the day by regular gunners of 506 Coast Regiment, Royal Artillery, and during the hours of darkness by the Home Guard.[18] This, of course, enabled significant savings in Royal Artillery manpower to be made.

Chapter 11

THE FINAL YEAR

January 1944 saw British troops engaged against the Japanese in Burma and against German forces in Italy. Italy had surrendered in September 1943 but German forces continued to defend strong positions in the country and there was no swift end in sight to the Italian campaign. In the east the Red Army crossed the border into Poland and the Germans were withdrawing on a wide front. The second front – the allied invasion of north-west Europe which would become known as Operation Overlord – could clearly not be long delayed. Britain was an armed camp with British, American, Canadian, Free French and Polish forces all training and preparing for the day when Hitler's western wall would be breached.

In the circumstances a German invasion of Britain was now hardly conceivable but the official view was that sabotage or diversionary raids were far from improbable and the Home Guard was regarded as a key element in the defence against such a contingency.

There had been for some time suggestions that the Home Guard was over-training its men, and that Guardsmen who had reached an acceptable degree of efficiency might be allowed to ease off training. This case had been argued by the Trades Union Congress in February 1943 and in that month it was announced that men who had reached a reasonable standard of military efficiency and who were engaged in work in the war industries and agriculture would have, as far as possible, calls on their time for Home Guard training reduced. It should be remembered that compulsory service had been introduced in February 1942, with a maximum compulsory commitment of 48 hours in four weeks, so a failure to attend parades or exercises laid a man open to possible criminal penalties. Even in the upper echelons of the Home Guard the occasional suggestion was made that fully trained men be given a 6-month leave of absence. Commenting on this, a diarist in the *Scotsman* observed: "After three years of regular parades, the original members of the Home Guard should be fairly well-versed in the art of defence. If they are not, then there must have been something radically amiss with their training."[1] The diarist made a fair point, although the situation was somewhat confused by the ever-changing role of the Home Guard and by the regular introduction of new weapons, both of which had considerable implications for training.

However there was a definite move towards a relaxation of training although the vital role that the Home Guard played continued to be emphasised. In May 1942, 1943 and 1944 there were special parades and ceremonies to mark the anniversary of the founding of the Local Defence Volunteers. The king and the prime minister issued statements and made speeches; Churchill even broadcast to the nation from the United States in May 1943 and the Home Guard mounted guard at Buckingham Palace on 14 May 1943. Representative parades were held in London with contingents travelling from all over the country to take part. This reflects the importance placed on the Home Guard in a period when unnecessary travel was frowned upon and every railway station had posters up asking "Is Your Journey Really Necessary?" Around Scotland, parades, weapons displays and demonstrations

had the dual function of allowing the Home Guard to show off
its qualities and letting the general public have the chance to man-
ifest support for their "people's army".

In May 1944 the king's fourth anniversary message to the
Home Guard clearly alluded to the coming invasion when he
spoke of: "a year when the duties assigned to you have a very special importance. To the tasks which lie ahead, the Home Guard
will be enabled to make a full contribution. I know that your
greatly improved efficiency, armament, and leadership render you
fit in every way for the discharge of these tasks."[2]

What these tasks now were was made clear to units by the
command structure. For example, in February 1944 Midlothian
Sector issued orders to its units outlining the most probable enemy threats. These were seen as the landing of small bodies of airborne troops with the aim of disrupting communications, gathering intelligence and destroying vital installations, and attacks by
stronger forces to immobilise ports and airfields with a view to
later exploitation of these facilities.[3] The Home Guard's defence
schemes had to provide for both these contingencies.

As regular troops began to move south to marshalling areas in
preparation for the invasion of Europe, the Home Guard found
itself with a new range of tasks to undertake. The Forth Bridge
had been guarded during the hours of darkness by a detachment
of the corps of military police but in June 1944 they were moved
away and the 2nd City of Edinburgh Battalion was ordered to
mount a picket of an NCO and six men at the south end of the
bridge from 8 p.m. to 6 a.m. Their comrades of the 7th Fife Battalion were assigned similar duties at the north end of the bridge,
and as that battalion's account points out: "Thus Home Guard
came full circle with the same guard duties as at the beginning
– the only difference was in the level of training and scale of
equipment."[4]

In Cambuslang, near Glasgow, a complex of telephone and telegraph premises including a building housing the Defence Telecoms Network had long been designated as vital points. Night guards were provided by two of the Glasgow Post Office battalions and on action stations the perimeter of the area would be guarded by a platoon drawn from the 5th Lanarkshire Battalion. In February 1944, with security being heightened in the run-up to the invasion, the protective cordon envisaged for an emergency was increased to two battle platoons of the 5th Lanarkshire, supplemented if needed by personnel from a nearby Z rocket battery.

Immediately before and after the Normandy landings in June 1944 Home Guard units across Scotland went onto a heightened level of readiness. The 3rd Perthshire Battalion, for example, arranged for one platoon to sleep each night at Battalion headquarters, armed and ready to move to any part of the area. This level of readiness was maintained until 13 July, after which date one platoon headquarters was manned each night so that a platoon could be quickly mustered to tackle any emergency.

As discussed in Chapter 9, anti-aircraft units continued to be developed and re-equipped through 1944 and change also took place in other fields. The 1st Midlothian Battalion, which had become too large for convenient command, was authorised in January 1944 to divide itself into two battalions. In Glasgow the 2nd City of Glasgow Battalion was busy in 1944 transferring 500 specially selected and trained men to its linked rocket battery. In Aberdeenshire the 4th (City of Aberdeen) Battalion formed a new support company with the battalion's anti-tank guns, sub-artillery and machine-gun platoons while the neighbouring 5th Aberdeenshire Battalion, based around Turriff, was asked to form a mobile company for operations, if needed, outside the battalion area.

All these changes and the introduction of new equipment

would provide some variety for the men of the Home Guard, and the heightened alert status around the time of the Normandy landing served as a welcome reminder that they still had an important role to play. However there was undoubtedly a degree of war-weariness and demotivation. The author of the account of the Dumfriesshire Zone Home Guard writes that: "it was becoming increasingly difficult to make the rank and file of the Home Guard realise that they would ever be called upon to play any part in forthcoming events. By the exertions of officers and NCOs, however, the men were kept together, but a sense of apathy – largely augmented by a cheap and irresponsible section of the Press – was clearly evident."

Other authors also speak of this problem, though one veteran of the 4th City of Glasgow Battalion remembered no difficulties with morale and attendance at drills and parades throughout the war[5] and the author of the 6th Lanarkshire Battalion's history noted that the majority of his unit's men, mostly working in the heavy engineering industry, maintained their enthusiasm to the end: "The greater number, however, stuck to their task and even in the last stages of the movement one could see bus loads after a day of field firing lasting all Sunday speeding back to Motherwell to change and go on to the evening shift."[6]

By September 1944 the German air threat had been reduced by the over-running by allied forces of the V-1 rocket launching sites in France and the Netherlands. There was now little risk of manned aircraft presenting a serious threat and the prospects of even German sabotage raids, let alone anything more serious, were vanishingly small. There remained only one German threat to the United Kingdom – V-2 rockets, which from September 1944 to March 1945 would be launched against the London area – but there was no anti-aircraft defence against this supersonic ballistic missile and no role for the Home Guard other than to assist Civil Defence in rescue work and clearing up after an attack. The Home Guard's role was ended, with the energies of its members to be better deployed in industry and other areas of civil society.

In September it was announced that compulsory parades would end on 11 September but that the Home Guard would remain in being as a voluntary service. Generally this made little difference and most men continued to turn out as the Home Guard reverted to the ideal of voluntary service which had distinguished its early years. However it was not clear what the future of the Home Guard was to be; would it continue on a voluntary basis after the war, and if not, how was it to be wound up? Indeed the end of the Home Guard came to have almost as much of an air of improvisation as had its formation. A company commander's letter in *The Times* spoke of regret that the four years of service were apparently going to fizzle out and resentment that the order for the discontinuation of compulsory parades should have come through a radio broadcast rather than through proper military channels.[7] The decision to end compulsory parades was, of course, sent out through channels but the secretary of state broadcast an announcement on the same night as the orders were promulgated – perhaps attempting to benefit from what political credit was to be had for the decision.

Practicalities had to be considered: what was to happen to the clothing and boots that had been issued? Lieutenant-Colonel W.R.P. Henry, writing from Duns, Berwickshire, suggested that men might be allowed to retain boots and greatcoats: "These items would be of immense value to the men, especially those living in country districts..."[8] In October it was announced that not only greatcoats and boots could be retained by men but battledress, anklets, badges, gas capes, respirators, and any spectacles, surgical appliances or dentures that had been provided by Home

Guard funds. This last concession by the War Office provoked the *Scotsman* to rather laborious editorial mirth:

> There were Home Guards who were not issued with artificial dentures. Small though the number may be, it is invidious that they should be denied, through no fault of their own, a parting gift bestowed on the many. One perhaps should not look a gift horse in the mouth, but the principle that one man should be given a toothsome morsel and another not is hardly equitable. If a Home Guard has not been issued with an artificial denture he should be given something else to make up for it. Say a bunch of War Savings certificates or a box of cigars.[9]

Leather belts, and leather jerkins (where these had been issued) were still needed overseas and as there was a shortage of suitable leather, these items were to be returned to stores, along with arms and military equipment. Greatcoats and boots of course were valuable items in a country where clothes rationing had been in force for years and "make do and mend" was the watchword. However it was emphasised that the Home Guard was not being disbanded, merely "stood down" with effect from 1 November 1944. Sir James Grigg, Secretary of State for War since 1942, told the House of Commons: "Members [of the Home Guard] are therefore liable for recall if the need arises. The instructions for standing down the Home Guard made it clear that should the Home Guard be recalled for duty, members will report for duty complete with the items of clothing and equipment which they have been allowed to retain. It will not be until the actual disbandment of the Home Guard that members will be allowed to dispose of, or have their khaki greatcoats, trousers, etc. dyed for use as civilian clothes."[10]

There was considerable interest shown in maintaining the

spirit of comradeship and some of the skills learned in the Home Guard. This chiefly manifested itself in the formation of Home Guard Rifle Clubs. In answer to a House of Commons question the secretary of state announced that County Territorial Army Associations would be asked to work out schemes suited to their area, and to control the loan of rifles and the issue of ammunition, and that such schemes should be in place by the stand-down date of 3 December.[11]

Compulsory parades may have ended on 11 September but the weeks that followed were not idle. Many men continued to parade and train on a voluntary basis and for storemen, adjutants, quartermasters, admin officers and indeed most officers and senior NCOs there was a huge job of preparing to gather in and return to depots all the various weapons, tools, equipment and retained clothing, and generally winding up an organisation which, in September 1944, had over 1.7 million men in its ranks.

How the service of these 1.7 million men (and approximately 30,000 women auxiliaries) was to be recognised became a matter of some controversy. Asked if the government planned to pay a gratuity to members of the Home Guard the secretary of state replied: "I am absolutely certain that it would be repugnant to the feeling of the vast majority of the Home Guard to ask for a gratuity." This was an opinion in which he was supported by the member for Ayr Burghs, Sir Thomas More, a Home Guard veteran, who asked: "Is the Minister aware that the Home Guard are proud to give their services free?"[12]

In the event those who served received a certificate bearing the king's signature, and recording the fact that the recipient, when the nation had been in mortal danger, gave of his time and was ready to defend the country by force of arms and with his life if need be. After the war another controversy broke out over the decision not to award the War Medal to the Home Guard; this had been awarded to all in the forces who served overseas, even if they never saw action, and it was pointed out that the contribution of the Home Guard, especially in areas such as anti-aircraft work, had been significant and merited such recognition.

Sunday 3 December was designated as the date for the final Home Guard parades. A Scottish contingent went south to join the national parade in Hyde Park, with three men being selected from every unit. One of the Scots who travelled south was Sergeant J.C. Smith of 12 Platoon, 3 Company, 1st West Lothian Battalion, who wrote an account of his trip to his battalion commander Lieutenant-Colonel Lord Charles Hope. On arrival in London they had been trucked to the Mill Hill Barracks where breakfast awaited them, orderlies were on hand to clean their equipment and a captain and a regimental sergeant major from the permanent staff at the barracks welcomed them. Smith wrote: "Once again it was most marked (as I have found on other occasions, courses, etc.) that the Regular troops did all in their power to honour the Home Guard."[13] A banquet for a randomly selected cross-section of the Home Guards, hosted by the Lord Mayor, was held at the Mansion House on the Saturday night, and on the Sunday Smith and 7,100 Home Guardsmen from across the country paraded in front of King George VI, the Queen and the Princesses Elizabeth and Margaret. Sergeant Smith noted: "One amusing sidelight was that the Colonel in charge at Mill Hill Barracks changed into civilian clothes, arranged to be near the Saluting Base when our contingent was passing, and gave us a special cheer." The Sergeant concluded his report by asking Lieutenant-Colonel Lord Hope to write to the colonel at Mill Hill thanking him for the way the Home Guard contingent had been treated.

While the London parade was the focus for the national commemoration of the work of the Home Guard, smaller parades took place across the country. The General Officer Commanding in Chief, Scotland, General Sir Andrew Thorne, took the salute at the Mound in Edinburgh as around 6,000 Home Guardsmen from Edinburgh, including men from the city's two rocket batteries and 24 coaches from the 1st Scottish Home Guard Transport Column, passed along Princes Street. In Glasgow nearly 10,000 Home Guards took part in the stand-down parade, marching past the Lord Provost, and a platform party consisting of Brigadier Hobart, the Glasgow sub district commander and the senior Home Guard officers of the city – the sub district Home Guard advisers, the sector commanders and the Post Office district commander.

Although participation in these parades was of course entirely voluntary, consistently high attendance figures were recorded across Scotland. The 1 Ross-shire Battalion, scattered over a large tract of northern Scotland and with a total strength of just over 1,700 all ranks, managed to turn out 740 men in pouring rain in Dingwall for its parade.

When the 3rd Aberdeenshire (South Aberdeenshire and Kincardineshire) Battalion paraded for the last time at Banchory, one of those who took part was 84-year old Charles Hunter, from the Stonehaven Company. He must have been 15 years too old to join the LDV back in 1940, but, as we have seen, the Home Guard was, to a considerable degree, a law unto itself. Across the land appropriate comments were made by provosts, lord lieutenants, senior officers and civic figures, but in a radio broadcast that night the king, who was not a great orator, summed up in what was considered to be one of his most effective speeches, the work and service of the Home Guard from its formation in a period of mortal danger when, "The most powerful army the world had ever seen had forced its way to within a few miles of our coast."[14] His majesty went on to recount the growth and development of the Home Guard and noted that: "For most of you – and, I must add, for your wives too – your service in the Home Guard has not been easy. I know what it has meant, especially for older men. Some of you have stood for long hours on the gun-sites, in desolate fields or wind-swept beaches. Many of you, after a long and hard day's work, scarcely had time for food before you changed into uniform for the evening parade."

After noting that the very existence of the Home Guard had helped ward off the danger of invasion and made possible the overseas campaigns of the army the king concluded with the thought: "But you have gained something for yourselves. You have discovered in yourselves new capabilities. You have found how men from all kinds of homes and many different occupations can work together in a great cause and how happy they can be with each other. That is a memory and a knowledge which may help us all in the many peace-time problems that we shall have to tackle before long."

The Home Guard was officially "stood down" with effect from 3 December 1944 but it continued in a curious administrative limbo for another year. Victory in Europe was celebrated in May 1945 and the defeat of Japan was achieved in August; throughout this time the Home Guard was, technically, still in being. Private Alfred F.B. Carpenter wrote to *The Times* in November 1945 pointing out that the invasion threat had diminished and that it was perhaps time to disband the Home Guard or release the 60 to 80-year-old group.[15] Private Carpenter (in another life a retired vice-admiral) was not the only person to hold such views, even if he expressed them with more wit than many. The Trades Union Congress, through its General Secretary, Walter Citrine, had earlier campaigned for a reduction in training commitments and had early in 1944 suggested that the Home Guard's role was over. The fear in the TUC now was that the Home Guard might be re-activated by the post-war Labour government to counter any problems caused by industrial disputes.

On 12 December 1945 the Labour government's Secretary of State for War, John Lawson, announced that the delay in formal disbandment had been to facilitate the recall of arms and equipment. While this was undoubtedly a factor in the delay there is also the possibility that the final disbandment was seen as something of a potential problem, the Home Guard having always been a rather difficult branch of the armed forces for politicians to deal with. In any event Lawson announced that the force would be disbanded with effect from 31 December 1945; all ranks would be deemed to be discharged and officers' commissions relinquished from that date, although officers with satisfactory service would be granted honorary rank in the highest rank they had held for an aggregate period of six months. Uniforms which had been retained now became the members' personal possessions and could be dyed or adapted as they pleased. Mr Lawson concluded by paying tribute to what he described as "a citizen force without parallel in our long history which sprang spontaneously into being to meet the greatest threat this country has ever faced."[16] Perhaps inevitably, certainly appropriately, the great Home Guard enthusiast, Sir Thomas Moore, Conservative member for Ayr Burghs, rose following the minister's statement to ask whether the Home Guard would receive a special medal. Lawson noted that they had received a certificate and would be eligible for the Defence Medal. This was awarded to those with three years' service in civil or military defence in the UK during the war.

What had been a large part of many lives for four years had ended; an event that understandably provoked mixed reactions. Most men were probably glad to be freed from the burden of service, however much they had enjoyed the comradeship and had, as the king had said, discovered new capabilities. Some had evidently enjoyed the break from domestic and work routine. The

Kirkintilloch Herald, announcing the disbandment of the Home Guard, hinted at one such reaction: "The Home Guard is to be disbanded on December 31, 1945 when members will cease to be liable to recall and the uniform which they have been authorised to retain will become their personal property, the War Office announces. A district man who has continued to give the Home Guard as his excuse for going out on certain evenings of the week will not welcome publication of the above."[17]

And so the Home Guard passed into the history books and into popular memory. So much of the current popular knowledge of the Home Guard is, as was pointed out in Chapter 1, coloured by the television series *Dad's Army*, that it is perhaps fitting here to emphasise that the phrase *Dad's Army* had no currency before the television programme of that name, and that we have a remarkable example of a fictional treatment of a historical event having almost totally coloured the perception of the event, not only in the minds of those too young to have memories of the war, but also in the minds of those who served in the Home Guard. This point is well made in Penny Summerfield and Corinna Pensiton-Bird's *Contesting Home Defence: Men, Women and the Home Guard in the Second World War*.[18] *Dad's Army* certainly caught on as a convenient popular term for the Home Guard, with its double sense of the army of our father's generation and an army of elderly men. Neither sense is of course entirely accurate but it is an undoubtedly useful and catchy title. Even a scholarly work like S.P. MacKenzie's *The Home Guard: a Military and Political History*, published by Oxford University Press, has on the cover of the paperback edition "The real story of *Dad's Army*". Whether the original working title for the series *The Fighting Tigers* would have had quite the same resonance and public uptake is rather questionable.

The Home Guard was actually revived, with an embarrassing lack of success, during the Cold War. The new Conservative government introduced a Home Guard Bill to parliament in November 1951; although this measure was included in their first Queen's Speech the plans for it had been developed by the previous Labour government. There was however neither the broad political support for the new force nor the public enthusiasm which had marked the formation of the LDV. The new force never reached its very modest manpower target and it was stood down in December 1955 and finally killed off with effect from 31 July 1957.

Chapter 12

SUMMING UP

If the *Dad's Army* image does not represent the whole truth about the Home Guard, how then should we see this remarkable mass movement which at stand-down had some 155,000 men in Scotland serving in battalions and batteries across the land? What did all this unpaid effort add up to? Was the Home Guard a serious force or just men playing at being soldiers? Was it all worthwhile? What, in short, was the Home Guard's contribution to winning the war?

Inevitably the search for answers to these questions is always going to be complicated by the fact that the Home Guard never met the enemy in the field, so there must always remain a doubt about what might have happened had a German attack or invasion taken place. That a determined and gallant resistance would have been mounted seems beyond question; how effective it would have been is of course another matter, and much would of course have depended on the nature of the enemy threat. One opinion was voiced by Bill McChlery, who served in the 4th City of Glasgow Battalion while working as a young man in Yarrow's Shipbuilders. He maintained that the Home Guard could have dealt very effectively with sabotage raids or commando-type raids, and that faced with a full-scale German invasion it could certainly have done some damage and annoyed the enemy; however on its own the Home Guard would have been incapable of doing more, simply because it wasn't trained or equipped to do more.[1] Major Lord Younger of the 1st Stirlingshire Battalion, reflecting on the problems of training and exercises and the fluctuating numbers of men available, nonetheless concluded: "Still, there were a sufficient number of keen men – particularly in some of the specialist jobs – to ensure that, if we had been embodied and had even a week or so to shake down, we should have been no undisciplined mob, but would have given a good account of ourselves."[2] Whether any likely crisis would have allowed the Home Guard the luxury of a week or so to shake down is, of course, highly questionable. Yes, perhaps a full-scale invasion would have seen the mobilisation of the Home Guard into a full-time force and the units more distant from the invasion sites would have had that time, but in the event of the more probable threat from sabotage or commando raids this breathing space would not have been available.

There may have been 155,000 Home Guardsmen armed and trained in Scotland in 1944 but it must be remembered that they were a dispersed and scattered force. Many battalions, simply due to geographical factors, never had the opportunity to exercise in anything larger than company-sized units and in many rural areas the only practical tactical unit was the platoon or the section. Quite how scattered a Home Guard battalion could be is suggested by the fact that the three battalions of the Kirkcudbrightshire Home Guard used 51 halls across the Stewartry for training purposes.[3] Faced with a German armoured force or even a substantial airborne attack it would have been folly to throw these amateur infantry units of the Home Guard into a mobile war situation when they lacked the training, command structures, the heavy weapons, the engineers, signals, services and other support to fight such a campaign with any hope of success. However that was never the function of the Home Guard, which had been set up as and remained a local defence organisation; its role was to take advantage of local knowledge and circumstances to locate, observe, fix, harry and delay an enemy, and buy the regular forces time to come to grips with the enemy. To compare the Home Guard with the regular army is not to compare like with like. On the other hand – as was noted in Chapter 5 – the Home Guard was, as much by accident as design, a very effective, if partial, response to the *blitzkrieg*. This was a mode of warfare where front lines were fluid and infiltration and attack in depth replaced the positional warfare of trenches that had characterised the First World War on the Western Front.

Major R.B. Sellar, a company commander in the 3rd Perthshire Battalion, wrote that the Home Guard had weaknesses when compared to the regulars, weaknesses arising inevitably out of their necessarily limited training. He identified particularly a reluctance to take cover and an inability to appreciate the importance of consolidating a position when it was taken: "In a recent operation, a company of amateur soldiers seized, by a really clever surprise attack, the enemy's transport. Half an hour afterwards when the regulars counter-attacked, they had very little difficulty in re-taking their vehicles. The Home Guards, flushed with victory, were playing cards."[4] To be fair, Sellar did acknowledge that these were largely the faults of past days; he was writing at the end of 1941 for publication in March 1942, and he considered that improved training and experience were going a long way to eliminate these problems.

What the Home Guard did, it did well. It provided a network of locally based armed response units able to deal with incidents like escaped German prisoners, and shoot down enemy aircrew. By its very presence in communities up and down the country it provided a degree of reassurance to the civilian population that there was some military force able to protect them. Thanks to its guide sections, with their knowledge of the local terrain, the Home Guard would be able in an emergency to help regular forces move swiftly and efficiently through its area and engage the enemy. In this context it should be remembered that regular forces were seldom stationed in an area long enough to allow them to get a good knowledge of the terrain; in an emergency the local expertise of the Home Guard guides would therefore have been invaluable. These guides were usually drawn from game-keepers, ghillies and farmers who were, by the nature of their daily work, intimately familiar with the local area and the best ways of moving men and equipment across it. By providing a garrison force the Home Guard also freed the regular army to train for and engage in its foreign war-fighting roles; as Major-General R.N. Stewart, who had been General Officer Commanding North Highland District wrote to the men of the 1st Moray Home Guard in 1945: "The army were able to make their great assault on Europe, leaving the base secure in your hands and knowing it would remain secure."[5]

One less obvious function of the Home Guard which is worth some thought is its role in training young men for the regular forces. Many young men joined the Home Guard while waiting call-up for the armed forces and received a sound train-ing in military skills. Although this training would need to be modified in the regular forces – with different rifles being used, for example – the value of the Home Guard training should not be overlooked. A young man like James Walker, who joined the 3rd City of Glasgow Battalion under-age and served with them for a year and a half was passed out by his battalion as proficient in use of the rifle, grenade and Sten gun, as well as in battlecraft and map-reading – all skills which would be emi-nently transferable to the regular army when he was called up and joined the 4th King's Own Scottish Borderers and fought with them in north-west Europe. Although Walker found the army equipment far superior to that which he had used in the Home Guard his basic military skills had been well learned before he put on the KOSB uniform.[6]

Very substantial numbers of men moved from Home Guard battalions to join the regular forces. The 2nd Lanarkshire Battalion, for example, whose average strength varied from 2,800 to 2,200 men, saw 850 members leave over its lifespan. This continual attrition of a Home Guard battalion had, naturally, consequences for its own internal training; however the flow of trained recruits to the armed forces must have been a significant contribution to the war effort. Some units, especially works units recruiting from essential services like the Post Office, railways, or defence industries, naturally supplied fewer men to the regular armed forces and enjoyed a greater degree of stability, but all to some degree took part in this training function. One other training role which came the Home Guard's way was work with young people in the revived Cadet Corps movement. Officers and NCOs from Home Guard battalions acted as instructors for the cadets and as examiners for their certificates.

An area where the Home Guard contribution must be assessed as being on a par with the regulars was their work in coastal artillery and anti-aircraft defence. Although the basic motivation in moving the Home Guard into these fields of work may have been to economise on regular manpower there seems to be no evidence that, when trained to operate coastal batteries, Z rockets, heavy or light anti-aircraft guns, they were not as effective in that role as the regulars. Although due to the timing of their move into anti-aircraft defence they had few opportunities to see active service, those opportunities that did come the way of the Home Guard rocket batteries were well taken and enemy casualties ensued. It is more difficult to assess the Home

Guard's work in coastal artillery, simply because, like the regular-manned batteries, the real-life challenge never came. However the evidence from reports of training exercises would suggest that the Home Guard batteries would have given a good account of themselves.

These artillery batteries were not, of course, the only areas where Home Guards provided a direct substitution for regular forces. The key road junction at Bonar Bridge, Sutherland, commanding the routes to the north and north-west, had, in the early years of the war, been designed to be defended by a battalion of regular troops. By early 1942 its defence was entrusted to the 1st Sutherland Battalion of the Home Guard, augmented by men of the Canadian Forestry Corps with a Home Guard officer as garrison commander; by June 1944 the 1st Sutherland had sole charge of this location. Similar replacement of garrison troops took place throughout the country. For example in 1943 the regular forces withdrew from Lunan Bay between Montrose and Arbroath. This large beach would have been a likely landing ground for any enemy attack and so the 1st Angus Battalion obtained approval to restructure itself to provide a mobile company to deal with beach landings and to create a headquarters company to house a small striking force to assist the mobile company. The striking force comprised an artillery platoon with two 18-pounder field guns, and a mobile infantry platoon providing escort to the guns and manning two medium machine-guns.

Despite what, in retrospect, appears to have been a low level of German threat, the reality of 1943 or 1944 was that there remained a perceived risk and it would have been quite impossible to mount and sustain major overseas campaigns in the Middle East, Far East, Italy and north-west Europe on the scale that Britain did if large regular forces had been required to remain behind in the United Kingdom on garrison duties.

One aspect of the Home Guard experience that is not often considered is the sheer physical strain that it put on those involved. The drills and parades, exercises and manoeuvres, all took energy and stamina, especially coming on top of long hours in civilian employment and the demands of life in a difficult wartime environment of shortages, air raids, rationing and general tension. This was alluded to by the Earl of Elgin, the Fife Zone Home Guard adviser, in his report on the zone in 1944: "On occasions it could be noticed that the strain of working hard all day and doing strenuous duties at night told its tale. . .at one parade of over 500 men, mostly country men at that, a look of under-nourishment was plainly to be seen, which seems to indicate that the civilian ration (for Home Guards at least) should have been of equal quantity to that given to the serving man, as those then serving in the County had a 'well-fed look' as compared to the Home Guard on parade."[7] While a basic diet was ensured to all who could afford it, through rationing, this was worked out scientifically on the basis of calories needed to sustain life in normal circumstances, and it did not, as the Earl of Elgin suggests, take appropriate account of the additional energy consumed in living the double life of worker and part-time soldier.

The suggestions made by the Trades Union Congress that the Home Guard was over-trained or indeed that the need for it had passed were very largely influenced by union officials' awareness of the strain under which men were working, although a certain anti-militarism and suspicion of the potential peace-time role of the Home Guard cannot be discounted. Trade unionists had memories of the armed forces being used in what they considered a strike-breaking role during industrial disputes between the wars and feared that a large force under military command might be used in future disputes.

Just as the Home Guard was moving towards stand-down, a

report published by a parliamentary select committee produced remarkable figures on what the force had cost. It found that its administration had been economically carried out and that it had been "efficiently but not extravagantly armed" and that the average cost for the past three years had been in the region of £9 per member.[8] In the current financial year, with a personnel strength of 1.8 million, the service would cost £16.6 million and the capital value of stores, clothing and equipment held by the Hone Guard came to almost £61 million. The principal revenue costs of the Home Guard were broken down as follows:

Grants to Territorial Army Associations for administration	£3.2m
Training schools, travelling instructors and permanent training staff	£2.2m
Subsistence	£1.6m
Travelling expenses and allowances	£1.3m
Clothing, general stores and personal equipment	£2.7m
Practice, ammunition and maintenance	£4.5m

With the annual subsistence cost for 1.8 million men totalling £1.6 million – or £0.89 per man – it can be seen that there were few occasions when Guardsmen qualified for subsistence payments. The report entirely confirms the view expressed two years earlier by Major R.J.B. Sellar that the Home Guard was "the cheapest army in the world".[9] Major Sellar had also commented that: "Almost every Home Guard has a civilian job, many of them very laborious ones. Yet these men who, in the best sense of the words, lead a double life, turn in a twinkling from civilian to soldier, taking the rigours of two careers in enthusiasm's easy stride."

Sellar, writing some years before the Earl of Elgin's comments, may have painted too rosy a picture but there is little doubt that there was an enormous enthusiasm and commitment which was sustained after the high threat period had ended. The surge of volunteers to join the LDV in May 1940 is perhaps not too surprising in light of the very real danger of German invasion, but the persistence of most of the members in attending training, parades and exercises after the invasion danger had receded is more remarkable; it should be remembered that the Home Guardsmen were made up of old soldiers from World War I who might be excused war-weariness and a bitterness that the war to end wars had been such a disappointment, as well as younger men – teenagers and men in reserved occupations in their 20s and 30s – who were of the generation which had voted at the Oxford Union in 1933 that "this House will in no circumstances fight for its King and Country." This commitment to service in the Home Guard may, in part, be explained by a sense of national unity in the face of the self-evidently evil force of Nazi Germany and the unquestioned threat to the nation's life. However it also reflects considerable credit on the leadership skills of the Home Guard officers, who, without the sanctions of military discipline to back them up, managed to raise, lead and motivate through more than four years what was an unpaid army and one which, after 1942, included significant number of non-voluntary members.

Sir James Grigg, who had made the unusual transition from being Permanent Under Secretary of State for War, the senior civil servant in the War Office, to becoming Member of Parliament for East Cardiff in 1942 and being appointed as Secretary of State for War, commented on the nature of the Home Guard when the end of compulsory parades was announced. (Sir James Grigg should not be confused with Sir Edward Grigg, MP, who had been Parliamentary Under Secretary of State for War in 1940 to 1942.) He said: "The truth is that the Home Guard parade was more than a bit of military training. It was an outward and visible sign of an inward unity and brotherhood, without distinction of class or calling, begotten of a great danger, continued until the danger had vanished, and we believe, to be carried forward into the future, when our task will be to rebuild our national life rather than to fight for its very maintenance."[10]

Many others made similar comments about the bonding and comradeship experienced in the Home Guard. The account of the 1st Battalion Scottish Borders Home Guard reflected in terms which were reminiscent of the king's stand-down broadcast and which considered the social consequences and the impact on personal growth and development engendered by the Home Guard.

A battalion of over 1,000, through which 3,000 men have passed, is a vast filter-bed, and men of all social groups have been thrown together for a number of years, with the happiest results. Men of like ideas have come together and men of opposite ideas have learned to respect those with which they do not agree; but running through the whole was that saving grace of humour, that ability not only to have a laugh at others but to laugh at themselves, which kept the whole show on an even keel throughout.

It is true that no one enjoyed the fun poked at the Home Guard more than the Home Guard themselves, but underlying seriousness was not impaired thereby.[11]

The Home Guard is undoubtedly a difficult organisation to sum up, in part because it was an ever-changing entity. It developed from the crisis-engendered Local Defence Volunteers, through a variety of roles from an anti-parachutist force, to a force manning road-blocks and fixed defences, to a more mobile force with battle platoons and counter-attack capability. Then, as the invasion threat declined, it went back to being a guard force which allowed the regular Army to concentrate on preparing to take the war to the enemy. The Home Guard never quite became what Tom Wintringham and his fellow International Brigaders hoped it might be: a people's militia which would both defend the nation against foreign threats and provide a guarantee against the risk of a sell-out by appeasers and Quislings. However it never became the organisation dominated by the "Colonel Blimp" faction that Orwell and the left feared either; yes, there were any number of retired generals and admirals in the Home Guard, but not all were in command positions. Furthermore, the Home Guard won for itself a considerable degree of autonomy and local independence, often to the acute annoyance of the military command structure and War Office bureaucrats.

This changing nature of the Home Guard is very evident with the benefit of 60 years of hindsight, but it was also very apparent to the men of the 1940s. The third anniversary parade in May 1943 in Edinburgh was headed by a detachment of men wearing the uniform and carrying the equipment of the LDV, an unspoken statement of how far the force had come. However, no matter what changes had taken place in men, in weapons, in tactics and in role, the Home Guardsman – the amateur soldier defending his home, his factory, his town or village, serving without pay and devoting a large part of his free time to learning the skills of the soldier – remained the constant factor in the "People's Army". Major J.D. Butler, second in command of the 7th Fife Battalion of the Home Guard, provides an excellent summation: "Only the men were the same and it mattered not at all whether they were the veterans who had made the Home Guard or the younger men who had been made by it. They ran true to type and produced a truly citizen army, the like of which had not been seen in history."[12]

NOTES

CHAPTER 1

1 *Picture Post*, 17 May 1941
2 *Hansard*, 22 May 1940 col. 260
3 Wintringham, Tom, *New Ways of War* (Harmondsworth, 1940)

CHAPTER 2

1 Amery, Leo S, *My Political Life*, Vol. 3 (London, 1955)
2 *The Times*, 8 April 1940
3 *Illustrated London News*, 13 September 1919
4 *The Times*, 25 July 1940.
5 *The Times*, 17 June 1940
6 *The Times*, 5 August 1940
7 *The Times*, 1 August 1940
8 National Library of Scotland (NLS), MS 3821
9 Slater, Hugh, *Home Guard for Victory!* (London, 1941)
10 *Scotsman*, 28 May 1940

CHAPTER 3

1 Eden, Anthony, *Freedom and Order* (London, 1947) pp.71-73

2   *Hansard,* 21 May 1940 col. 8, 9

3   NLS MS 3821

4   *Hansard,* 19 November 1940 col. 1928-1932

5   *Scotsman,* 21 May 1940

6   *Hansard,* 17 September 1940 col. 86,87

7   NLS MS 3818

8   Hansard, 22 May 1940 col. 261

9   *Hansard,* 4 June 1940 col. 469

10 War Office Telegram Ref 8560 (A.G.1.A) 15/5

11 NLS MS 3821

12 War Office Letter 27/Gen.2594 (A.G.1.A)

13 *Hansard,* 10 November 1940 col.1941

14 NLS MS 3818

15 *Scotsman,* 25 May 1940

16 *Picture Post,* 29 June 1940

17 *Scotsman,* 17 May 1940

18 *Hansard,* 20 Aug 1940 Col.1095, 96

19 War Office Letter 27/Gen.2594 (A.G.1.A)

20 NLS MS 3816-3822

21 NLS MS 3820

22 NLS MS 3820

23 NLS MS 3818

24 *Hansard,* 22 May 1940 col. 246

25 NLS MS 3821

CHAPTER 4

1   NLS MS 3817

2   Graves, Charles, *The Home Guard of Britain* (London, 1943) p.70

3   Graves *op. cit.* p.71

4   Wintringham, Tom, *New Ways of War* (Harmondsworth, 1940)

5   *Hansard*, 4 June 1940 col. 796
6   Churchill to Eden 26 June 1940 quoted in Gilbert, Martin, *The Churchill War Papers* Vol 2. (London, 1995) p. 422
7   *The Times*, 15 July 1940
8   Pownall Diaries, 29 July 1940 (Liddell Hart Centre for Military Archives) quoted in Mackenzie, S.P., *The Home Guard* (Oxford, 1995) p. 49
9   *Hansard*, 23 July 1940 col. 576
10 *The Times*, 16 July 1940

CHAPTER 5

1   NLS MS 3816
2   National Archives WO199/2684
3   NLS MS 3818
4   Sillitoe, Sir Percy, *Cloak Without Dagger* (London, 1956)
5   *Glasgow Herald*, 31 July 1940
6   Notes on duties of Patrols and Sentries. . . n.d. in Stirling Council Archives PD43/1
7   Register of Deaths, Fortingall, Perthshire: Register of Corrected Entries 3 Sept 1940
8   *Scotsman*, 18 July 1942
9   *Scotsman*, 10 October 1944
10 NLS MS 3821
11 Angus Archives MS138
12 Borders Archives SC/R/48/1
13 NLS MS 3818
14 NLS MS 3820
15 *Scotsman*, 27 July 1940
16 Wintringham, Tom, *op. cit.*
17 National Archives WO32/10016
18 NLS MS3821
19 *The Scots Magazine*, March 1942

20 *Scotsman*, 19 October 1940
21 National Archives WO32/10616
22 *Hansard*, 6 November 1940 col. 1351
23 *Tribune*, 20 December 1940 p.8
24 *Hansard*, 19 November 1940 col.1891
25 National Archives WO32/14936
26 *Scotsman*, 12 September 1940
27 Angus Archives MS138
28 *Glasgow Herald*, 17 October 1940
29 *Glasgow Herald*, 22 October 1940
30 NLS MS 3818
31 NLS MS 3821
32 *Hansard*, 19 November 1940 col.1889
33 NLS MS 3819
34 Brophy, John *Home Guard, a handbook for the LDV.* 1940.
35 Brophy, *op. cit.*
36 *The Scots Magazine*, March 1942
37 NLS MS 3819
38 Angus Archives MS138
39 *Hansard*, 19 November 1940 col. 1950
40 *Hansard*, 19 November 1940 col. 1951

CHAPTER 6

1 National Archives WO199/2684
2 NLS MS 3821
3 Brophy, John: *Home Guard Proficiency* [1941] p. 70
4 NLS MS 3819
5 Home Guard Instruction, No. 51, Battlecraft and battle drill for the Home Guard. *The Organisation of Home Guard Defence* Part IV (1943) pp.15-16
6 *op.cit.* p.16

7   *op.cit.*, p.17
8   National Archives WO199/868
9   National Archives WO199/868
10 *op.cit.*, p.18
11 NLS MS 3819
12 NLS MS 3820
13 *Hansard*, 17 March 1942 col 274/5
14 *Hansard, op.cit.* col. 295
15 *Scotsman*, 12 February 1942
16 NLS MS 3821

CHAPTER 7

1   *Glasgow Herald*, 5 July 1943
2   NLS MS 3819
3   *The Times*, 15 August 1941
4   NLS MS 3819
5   National Archives WO32/10011
6   Angus Archives MS138
7   NLS MS 3819
8   *Glasgow Herald*, 28 July 1941
9   *Glasgow Herald*, 1 June 1942
10 *Scotsman*, 10 June 1942
11 *Scotsman*, 16 May 1942
12 NLS MS 3818
13 NLS MS 3818
14 Bill McChlery interview, 5 February 2008
15 NLS MS 3821
16 NLS MS 3819
17 *Scotsman*, 30 July 1941
18 *Scotsman*, 8 March 1943
19 NLS MS 3820
20 *Scotsman*, 23 January 1941

21 *Sunday Post*, 7 December 1941
22 Boyd-Orr, Lord, *As I Recall* (London, 1966) p. 140
23 NLS MS 3820
24 I am grateful to Peter Bilbrough for details of Private Halloran
25 *Scotsman*, 6 April 1943
26 *London Gazette*, 9 July 1943. I am indebted for this Gazette reference to Peter Bilbrough
27 NLS MS 3819
28 NLS MS 3821
29 NLS MS 3821
30 NLS MS 3819
31 NLS MS 3817
32 NLS MS 3816
33 National Archives WO199/2785
34 Stirling Archives PD43
35 National Archives WO199/2846
36 NLS MS 3818
37 NLS MS 3818
38 NLS MS 3818
39 NLS MS 3819
40 NLS MS 3821
41 Mackenzie, Compton, *My Life and Times 1939-46 Octave 8* (London, 1969)
42 NLS MS 3821
43 National Archives WO199/2684
44 *The Times*, 7 March 1941
45 *Hansard*, 10 December 1941 col.251/2
46 *Hansard*,17 March 1942 col.284
47 *The Times*, 18 December 1941
48 Hansard (House of Commons), 18 December 1941 col. 2130

49 NLS MS 3818
50 *Hansard,* 17 November 1942 col.65-6
51 *Scotsman,* 19 January 1943
52 NLS MS 3819
53 *Scotsman,* 15 September 1942
54 *Scotsman,* 16 September 1942
55 NLS MS 3820
56 Bill McChlery interview, 5 February 2008
57 Stirling Archives PD43
58 NLS MS 3816
59 *The Times,* 12 November 1943

CHAPTER 8

1 National Archives WO199/3288A
2 *Hansard,* 22 May 1941

CHAPTER 9

1 *Glasgow Herald,* 3 April 1942
2 *Scotsman,* 29 November 1943
3 "The Anti-Aircraft Defence of the United Kingdom." Supplement to the *London Gazette,* 16 December 1947
4 National Archives WO32/9753
5 National Archives CAB120/363
6 National Archives WO199/2763
7 *Glasgow Herald,* 1 April 1943
8 NLS MS 3818
9 National Archives WO199/2787
10 National Archives WO 199/2787
11 *Edinburgh Evening Dispatch,* 25 September 1944
12 NLS MS 3822
13 National Archives WO199/2756
14 National Archives WO199/2756

15 "The Anti-Aircraft Defence of the United Kingdom." Supplement to the *London Gazette*, 16 December 1947

CHAPTER 10

1   NLS MS 3820
2   National Archives WO199/346
3   National Archives WO199/2814
4   "Gordon Highlander", 4th LNER Battalion, HQ Coy, Home Guard. n.d.
5   National Archives WO199/2814
6   National Archives WO199/2814
7   NLS MS 3820
8   *Scotsman*, 21 February 1944
9   NLS MS 3821
10 National Archives WO199/3388
11 National Archives WO199/3251
12 National Archives WO199/2892
13 National Archives WO199/3251
14 National Archives WO199/2892
15 National Archives WO199/2892
16 National Archives W0199/2684
17 *Scotsman* 13 October 1943
18 NLS MS3821

CHAPTER 11

1   *Scotsman*, 17 September 1943
2   *Scotsman*, 15 May 1944
3   NLS MS 3820
4   NLS MS3821
5   Bill McChlery interview, 5 February 2008
6   NLS MS 3818
7   *The Times*, 18 September 1944

8   *The Times*, 27 September 1944
9   *Scotsman*, 13 October 1944
10   *Hansard*, 14 November 1944 col. 1785
11   *Hansard*, 28 November 1944 col. 2379-80
12   *Hansard*, 14 November 1944 col. 1786
13   NLS MS 3821
14   *The Times*, 4 December 1944
15   *The Times*, 17 November 1945
16   *The Times*, 13 December 1945
17   *Kirkintilloch Herald*, 19 December 1945
18   Summerfield, Penny and Peniston-Bord, Corinna, *Contesting Home Defence: Men, Women and the Home Guard in the Second World War* (Manchester, 2007)

CHAPTER 12

1   Interview with Bill McChlery, Kirkintilloch 5 February 2008
2   NLS MS 3819
3   NLS MS 3819
4   *The Scots Magazine*, March 1942
5   Boyd, A, *A Brief History of the 1st Moray Battalion Home Guard* [Elgin? nd.] Foreword
6   Interview with James Walker, 4 January 2008
7   NLS MS 3821
8   *Glasgow Herald*, 22 November 1944
9   *The Scots Magazine*, March 1942
10   Quoted in Lauriston *1940-44: The Story of A Company of the 8th Battalion, City of Edinburgh Home Guard* (Edinburgh, 1945)
11   NLS MS 3820
12   NLS MS 3821

INDEX

Note: Within headings 'HG' has been used for Home Guard and its predecessor Local Defence Volunteers